AF886212

Medical Use of Human Beings

Whilst activities like transplantation and medical research have typically been considered on a discrete basis, they are also actually part of a broader phenomenon of medical means being employed to make use of human beings. This book is the first ever systematic critique of such medical use of the human being as a whole. It is divided into two parts. The first part considers what constitutes an appropriate normative lens through which to view such medical use and its constraint. It makes a reasoned ethical and human-rights-based case for preferring respect for human worth over any of the main alternative approaches that have been drawn on in specific contexts and outlines what this preference practically implies. The second part uses this respect-based lens to critique use discourse, law and practice. Drawing on three contrasting case study areas of warfare-related medical use, transplantation and human tissue research, this book exposes both the context-specific and thematic nature of shortfalls in respect.

Overall this book provides a compelling analysis of how medical use ought to be constrained and a compelling critique of the excesses of discourse, practice and governance. It is recommended to academics, students, policymakers and professionals whose work is focused on or intersects with the medical sector and anyone else with an interest in medicine and its limits.

Dr Austen Garwood-Gowers is a reader in health law and ethics at Nottingham Trent University. He is author of over 60 publications in this field, having been a dedicated teacher and scholar since the early 1990s. His more recent work has brought a new understanding of fields like transplantation and medical research, showing that they are not just interconnected but also elements of a broader phenomenon of medical means being employed to make use of human beings. He breaks more new ground in this book by subjecting this broader phenomenon of medical use of human beings to its first ever systematic critique.

Biomedical Law and Ethics Library

Scientific and clinical advances, social and political developments and the impact of healthcare on our lives raise profound ethical and legal questions. Medical law and ethics have become central to our understanding of these problems, and are important tools for the analysis and resolution of problems – real or imagined.

In this series, scholars at the forefront of biomedical law and ethics contribute to the debates in this area, with accessible, thought-provoking, and sometimes controversial, ideas. Each book in the series develops an independent hypothesis and argues cogently for a particular position. One of the major contributions of this series is the extent to which both law and ethics are utilised in the content of the books, and the shape of the series itself.

The books in this series are analytical, with a key target audience of lawyers, doctors, nurses, and the intelligent lay public.

Series Editor: Sheila A. M. McLean
Professor Sheila A.M. McLean is Professor Emerita of Law and Ethics in Medicine, School of Law, University of Glasgow, UK.

Available titles:

Healthcare Ethics, Law and Professionalism
Essays on the Works of Alastair V. Campbell
Edited by Voo Teck Chuan, Richard Huxtable and Nicola Peart

Religion, Medicine and the Law
Clayton Ó Néill

Medical Use of Human Beings
Respect as a Basis for Critique of Discourse, Law and Practice
Austen Garwood-Gowers

For more information about this series, please visit:
www.routledge.com/Biomedical-Law-and-Ethics-Library/book-series/CAV5

Medical Use of Human Beings

Respect as a Basis for Critique of Discourse, Law and Practice

Austen Garwood-Gowers

Routledge
Taylor & Francis Group
LONDON AND NEW YORK

First published 2020
by Routledge
2 Park Square, Milton Park, Abingdon, Oxon OX14 4RN

and by Routledge
52 Vanderbilt Avenue, New York, NY 10017

Routledge is an imprint of the Taylor & Francis Group, an informa business

© 2020 Austen Garwood-Gowers

The right of Austen Garwood-Gowers to be identified as author of this work has been asserted by him in accordance with sections 77 and 78 of the Copyright, Designs and Patents Act 1988.

All rights reserved. No part of this book may be reprinted or reproduced or utilised in any form or by any electronic, mechanical, or other means, now known or hereafter invented, including photocopying and recording, or in any information storage or retrieval system, without permission in writing from the publishers.

Trademark notice: Product or corporate names may be trademarks or registered trademarks, and are used only for identification and explanation without intent to infringe.

British Library Cataloguing-in-Publication Data
A catalogue record for this book is available from the British Library

Library of Congress Cataloging-in-Publication Data
Names: Garwood-Gowers, Austen, author.
Title: Medical use of human beings : respect as a basis for critique of discourse, law and practice / Austen Garwood-Gowers.
Description: Abingdon, Oxon ; New York, NY : Routledge, 2020. | Includes index.
Identifiers: LCCN 2019016222 | ISBN 9781315605814 (eb)
Subjects: LCSH: Human experimentation in medicine—Law and legislation. | Human body—Law and legislation.
Classification: LCC K3611.H86 G375 2020 | DDC 344.04/196—dc23
LC record available at https://lccn.loc.gov/2019016222

ISBN: 978-0-754-67964-6 (hbk)
ISBN: 978-1-315-60581-4 (ebk)

Typeset in Galliard
by Apex CoVantage, LLC

Contents

Figures

Table

Abbreviations

Banjul Charter	African Charter on Human and People's Rights, 1981
CHRB	Convention on Human Rights and Biomedicine, 1997
ECHR	European Convention on Human Rights, 1950
ECtHR	European Court of Human Rights
HRA	Human Rights Act, 1998
PVS	Persistent Vegetative State
UDHR	Universal Declaration of Human Rights, 1948
UDBHR	The Universal Declaration on Bioethics and Human Rights, 2005
UDHGHR	Universal Declaration on the Human Genome and Human Rights, 1997

Introduction

The context to me writing this book

In the 1990s I became heavily focused on the ethical and legal aspects of living donor organ transplantation, gaining a PhD from De Montfort University on the subject and authoring a number of publications including a groundbreaking monograph entitled *Living Donor Organ Transplantation: Key Legal and Ethical Issues* (Garwood-Gowers, 1999). By the late 2000s I was beginning to think in much more detail about the links between different forms of medical practice that were either less than optimally therapeutic or non-therapeutic. I began to perceive a pattern of dilution of protection of human beings across these areas being not just advocated at times in the discourse but also reflected in practice and at times even permitted by governance and law. I wrote a chapter entitled 'The right to bodily security vis-à-vis the needs of others' for the edited collection *Autonomy and Human Rights in Healthcare* (Garwood-Gowers, 2008) which exposed this and began to make the normative case against it. Since then I have more explicitly refined my focus to the medical context and recognized that what I am essentially reflecting on is a phenomenon of use of human beings in this context, or what I initially described as medical use of the human body and have now come to describe in this book as medical use of the human being. This phenomenon may be longstanding, but it has not been overtly recognized as a phenomenon before my work in this area and consequently never actually systematically addressed as such. The key reason it needs to be is that discourse, practice, governance and law that is specific to different forms of use (such as in contexts like transplantation, general medical research, human tissue research and so on) is 'out of whack' with the *overarching* norms of health law, human rights and, I will argue, ethics that they ought to be conforming to. And the key reason why this disjuncture is 'screaming' to be written about in detail is that it is ultimately a problem of lack of respect for human worth. The norms themselves point to the need for such respect but to a very significant extent use context specific discourse, practice and to some extent even governance and law reflects an agenda of diluting emphasis on respect in order to bring about greater use. I have found the disjuncture not just interesting but disturbing. That turned writing this book into something of a mission. This was a blessing since, whilst I generally find writing a joy, this is the most intellectually rich and multifaceted piece of work I have created, and there were some moments of difficulty and confusion when it became like the emotional equivalent of pulling one's teeth out.

The phenomenon of medical use of the human being

It was only a short and obvious jump from talking about the use of human beings in the medical context to coining the phrases 'medical use of the human body' and 'medical use of the human being.' Both phrases capture the phenomenon of practices that fulfil the criteria of:

(1) Intruding on human interests by their performance at least partly to achieve 'wider' ends; *and*
(2) Being medical in character such as by entailing the use of medical knowledge, technology or intervention.

The intrusion on interests will typically involve some kind of intervention on the body. An illustrative example would be cloning of human embryos for research purposes. Typically to achieve wider research ends, this impinges on the physical security of those embryos on which it is performed, interferes with their form and with the form of humans as a species. However, it is also worth noting that some intrusions are purely mental in character, and this is why I settled on preferring the phrase 'medical use of the human being' rather than the phrase 'medical use of the human body'.

An example of purely mental intrusion is the use of psychological techniques to bring about mental disintegration and degradation. Many law enforcement agencies routinely use some of these techniques to extract information and confessions out of suspects. They have also been used by a large range of states, for example, against political dissidents during periods of repression and against suspected enemy combatants during times of armed internal conflict and war. In the latter context, many readers will be familiar with their use in prison camps like Guantanamo as part of the USA's so-called War on Terror (see further Moreno, 2012, 77–97). Another example is the so-called twins study conducted in the 1960s and 1970s by the child psychiatrist and psychoanalyst Peter Neubauer. The study involved an unknown number of sets of identical Jewish twins and triplets given up to Louise Wise Adoption Services for placement with Jewish families in the New York area. The agency organized placements in a manner that was contrived to fit the to-this-day still partially hidden purposes of the study rather than to maximize the protection of the interests of the children involved. No attempt was made to keep each set whole, seemingly because Neubauer wanted to study the effect of family background and other environmental factors on childrearing. The families the children were placed with were not informed of the study, the fact that this was the real reason why they were being visited by researchers on an ongoing basis or even the fact that the children were part of a twin or triplet pair. Some of them have subsequently found out about their (mis)use by accident, but even they have struggled to get information, and there are shocking plans to keep the records of the study sealed at Yale University until 2065 (McCormack, 2018). Several of the children (mis)used went on to commit suicide as adults, raising potential questions about both the damage done to them by being separated, potentially by design placed in families within which parenting issues were expected to arise and then secretly studied on an ongoing basis. It is strangely and disturbingly ironic that an Austrian Jew who had fled the Nazis should develop and run a study with a Jewish adoption agency that involved almost Nazi style abuse of Jewish twins.

The second criterion embodies an inclusive approach to what qualifies as medical when medical use is being referred to. This was a deliberate attempt on my part to capture the 'warts and all' reality of how relevant discourse, practice, governance and law approach uses of the human being that could be described as in some significant way 'medical in character,' rather than a partial and sanitized view of it. To that end as well it is notable that the second criterion allows for a use to be classified as medical where it is brought about by medical means even when the ends it pursues are not related to medicine or are actually antithetical to it. Thus, for example, I classify the warfare related testing of chemical, biological and radiological agents and weapons and the construction of DNA databases for criminal justice purposes as being just as much forms of medical use as transplantation.

Breaking new ground with a holistic consideration of medical use

Medical use of the human being is not just a novel phrase but also a description of a phenomenon that has hitherto not been holistically examined. I alluded to this earlier, but it is worth reflecting on it here in more detail. Previously, the interrelated fields of medical, health and bioethics and law have tended to examine specific forms of medical use as if they were best treated as discrete phenomena when in fact they are all parts of a singular whole. General textbooks in the field, for example, have typically devoted a chapter to each of what are considered to be the most important forms of use such as medical research and transplantation. They have, at most, only ventured beyond this 'silo' approach to something more cross-cutting when dealing with the treatment of the dead. Specialist books are little better in as much as they tend to nearly always be focused on one form of use (cf. two in the case of Price, 2010) or to cut across forms but only to the end of dealing with one issue or theme such as commodification or one class such as the dead (e.g. Jones & Whitaker, 2009). *The Body Divided: Human Beings and Human 'Material' in Modern Medical History* (eds. Wilde & Ferber, 2016) is perhaps something of an exception, but as an edited collection, its represents not so much an attempt to forge a theoretical perspective on medical use as a whole but a detailed historical analysis of different forms of use – and despite being eclectic in its range of chapter is very heavily focused on dissection and anatomy. Some legal regimes have a single piece of legislation that addresses at least most forms of medical use of extracted material within it. However, such legislation does not extend to also covering forms of medical use that do not entail extraction, save the fact that in some cases it will cover use of the whole bodies of the dead.

Considering medical use in such partial ways has its place, not least in legislation where it allows for greater specificity of treatment and guards against the worsening of already acute problems of unwieldiness. However, it also has its limitations. Not least of these is that it makes it more difficult to pick out and understand issues that are common across parts it has been divided into – such as common across field and class of human used. This is significant not just scholastically and practically but also ideologically. Specifically, it has enabled the dilution of protection agenda to be advanced in a manner that is class and/or context specific with little reference to its wider implications, giving the impression that it is somehow exceptional and that the problems raised by it can thereby be contained. A more holistic presentation of the discourse, practice, governance and law pertaining to medical use is vital because it exposes this notion to be fallacious. Let me give a few examples of this. If one makes the argument that one should dilute protection of the individual with respect to post-mortem use of their organs for transplantation one could potentially just as well be making for other forms of post-mortem use that are conventionally regarded as needing to be expanded. If one argues that the best interests of incapable adults should not be protected in medical research should not be protected, one could equally argue the same in such other contexts as transplantation and also argue it for at least incompetent minors. If one argues that informed consent is not needed for the use of human materials and data derived from them in tissue research, one could equally argue that it should be abandoned in other use contexts. Both the appetite for medical use and for strands of thought that are incompatible with respect are strong and operate in combination to push for dilution in a dynamic way across a range of contexts.

How medical use is a generally problematic phenomenon

Notwithstanding what I have already said, I am not suggesting that medical use is inherently problematic. Indeed, some forms of it are an almost natural element of the routine practice of medicine within any given system. For example, when a human being uses medical intervention to try and heal or ameliorate the condition of another, it is natural that they might use that experience to learn from the results of doing so and potentially to pass that learning on to others. However, taken as a whole medical use is very open to being brought about both for problematic ends and by problematic means.

In terms of ends, arguably the greatest issue comes with forms of use which are antithetical to medical ideals. Warfare related medical use, discussed in Chapter 4, often fits into this category as much of it is aimed at destroying rather than healing human beings. However, there is also a more general issue of the productivity of medical use being overstated. Not surprisingly, most people write and think about the value or otherwise of specific forms of use from the perspective of prevalent orthodoxy concerning what constitutes 'good medicine.' Yet it is doubtful that prevalent thinking about what constitutes good medicine has ever fully corresponded with what good medicine *actually* is. One of the themes of this book is that we should embrace this fact by questioning rather than colluding with orthodoxy and forms of medical use grounded on it.

This is a task I take to with some gusto in Chapter 1. I make the case that the mechanistic orientation of establishment medicine goes against the paradigmatic orientation of the hard sciences on which it claims to be based. This, combined with its industrialized character, has rendered it counterproductive. The material effects of this are felt across a range of forms of medical use that are conventionally thought to be of value, including solid organ transplantation and human tissue research, which respectively form the focus of Chapters 5 and 6. Transplantation is shown to be in various respects inherently problematic whilst human tissue research is shown to be practiced in ways that reflect and foster a mechanistic and industrialized approach rather than help us move beyond it. Notwithstanding radical theories of liberty, allowing destructive, unproductive or counterproductive use to be carried out on even a knowing and willing rational agent (such as a capable adult) is problematic. It is not only anti-social and uneconomic but also inconsistent with the theory of respect for human worth that I advance in detail in Chapter 1.

In terms of the *means* by which medical use is pursued, problems start to arise when optimal protection of the therapeutic and/or other interests of those being intervened on is departed from. The tendency to make such departures has become normal and extensive in the modern era both with respect to use of the whole body and parts extracted from it. This can be related to the way resistance to use of the body has withered over time. Consider dissection of the bodies of the dead for research and educational purposes. In ancient Greece, it was a somewhat taboo practice. Aristotle only performed it on animals and human foetuses. Herophilus (350–280 BC) went further and used dead human beings, but these were mostly condemned criminals. In Europe, resistance to its use continued in the medieval era in spite of the fact that the Christian cult of relics had normalized the practice of dismemberment (Park, 1995). However, over time that resistance has been broken down. By the middle of the nineteenth century, it was being performed extensively on the poor in Great Britain (MacDonald, 2016) and parts of its empire (Martin, 2016). And today it is practiced on a mass scale and generally only becomes the subject of societal opprobrium when practiced incompatibly with expected norms, such as without consent. Consider also the broadly positive reception that has been given to such practices as transplantation and biobanking; something that would have been unthinkable in earlier eras.

Creating a lens or benchmark through which to critique medical use

Whilst this book is not directed at elucidating the *precise circumstances* in which any given form of medical use is unethical, the first two of three chapters that form Part I of it do reach conclusions on what, as a matter of ethics, constitutes an appropriate *approach* to its constraint. Chapter 1 starts this process by exploring the nature and implications of a respect based approach to ethics both generally and then more specifically in the context of medical use. In it I illustrate the clash between limited and more inclusive approaches to understanding and protecting the worth of human beings individually and as a species. I make a case for the latter, grounded on a broader understanding of the nature and purpose of reality that encapsulates within it a concept of human flourishing. I explore the specific implications of this approach for medical use of the human being, asserting that, in order for such use to protect rather than degrade or devalue human worth, it must be both productive in terms of the ends it pursues and compatible with certain norms in terms of the means by which it pursues them. I assert that these norms consist of:

(1) Protecting all members of the species as of equal inherent value;
(2) Protecting human nature and form;
(3) Protecting the species and its members from instrumentalization; and
(4) appropriately accounting for relevant non-absolute principles, such as autonomy and beneficence.

In Chapter 2, I explore the principal alternative theoretical benchmarks that have been used to dilute protection in the context of medical use. These are libertarianism and certain approaches to justice and utilitarianism. Ultimately, I argue that the case that has been made for each of these is weak.

Chapter 3 explores where human rights norms stand in relation to this issue. Some ethicists in the discourse have suggested that it is normal for legal and other rules to allow for the needs of society to be prioritized over the bodily security of human beings. For example, Calabresi (1991, 12) cites rules restricting the mutilation of our bodies and selling our body parts as inconsistent, and Harris cites 'control of dangerous drugs, control of road traffic, vaccination, screening tests, blood donation, quarantine for communicable disease, compulsory military service, detention under mental health acts, restriction of sexual activities and professional activities of HIV positive people' (1997, 12). However, these examples are actually more precisely understood as restrictions of liberty designed either to protect those subject to them and/or the conflicting rights of others. Nonetheless, there are a considerable number of legal and other rules that, if viewed in isolation, permit protection of the interests of the individual to be diluted to meet the mere (perceived) needs of others. However, the phrase *taken in isolation* is critical here. The moral, political and often even legal framework in this as so many other contexts is set by human rights frameworks and instruments – particularly at the global and regional levels. And, as I illustrate in this chapter, they contain a range of general norms and specific protections that sit at odds with diluting protection to meet (perceived) need.

References

Calabresi, G. (1991) Do we own our own bodies. *Health Matrix.* 1(5), 5–18.

Garwood-Gowers, A. (1999) *Living Donor Organ Transplantation: Key Legal and Ethical Issues.* Aldershot: Ashgate.

Garwood-Gowers, A. (2008) The right to bodily security vis-à-vis the needs of others. In Weisstub, D. & Diaz-Pintos, G. (eds.) *Autonomy and Human Rights in Healthcare: An International Perspective.* London: Springer, pp. 375–397.

Harris, J. (1997) The ethics of clinical research with cognitively impaired subjects. *Italian Journal of Neurological Sciences.* Supplement, 5, 9–13.

Jones, G. & Whitaker, M. (2009) *Speaking for the Dead.* Aldershot: Ashgate.

MacDonald, H. (2016) A body buried is a body wasted: The spoils of human dissection. In Wilde, S. & Ferber, S. (eds.) *The Body Divided: Human Beings and Human 'Material' in Modern Medical History.* Abingdon: Routledge, pp. 9–28.

Martin, S. (2016) Dissection, anatomy acts and the appropriation of bodies in nineteenth century Australia: 'The government's brains' and the benevolent system. In Wilde, S. & Ferber, S. (eds.) *The Body Divided: Human Beings and Human 'Material' in Modern Medical History.* Abingdon: Routledge, pp. 53–74.

McCormack, W. (2018) Records from controversial twin study sealed at Yale until 2065. *Yale News,* 1 October. Available at https://yaledailynews.com/blog/2018/10/01/records-from-controversial-twin-study-sealed-at-yale-until-2065. Accessed 5.3.2019.

Moreno, J. (2012) *Mind Wars.* New York, NY: Bellevue Literary Press.

Park, K. (1995) The life of the corpse: Division and dissection in late medieval Europe. *Journal of the History of Medicine and Allied Sciences.* 50, 111–132.

Price, D. (2010) *Human Tissue in Transplantation and Research: A Model Legal and Ethical Donation Framework.* Cambridge: Cambridge University Press.

Wilde, S. & Ferber, S. (2016) *The Body Divided: Human Beings and Human 'Material' in Modern Medical History.* Abingdon: Routledge.

Part I

The case for respect in the context of medical use of the human being

1 The ethical case for respect as a basis for constraint

1.1. Introduction

Respect is not just a principle or concept but also an attitude (Downie & Telfer, 1969) of unconditional positive regard for whoever and whatever possesses worth. However, for millennia there has been a contestation of whether worth is possessed by all aspects of reality and reality as a whole or is possessed only or more heavily by certain aspects of reality – such as human beings or only a subset of them who satisfy 'qualifying' criteria. In Western discourse what amount to tussles over this have often been played out through the discourses relating to personhood and dignity – with the application of the former being necessarily narrow and the latter often interpreted as being so but at least capable of being broader. Having critically explored how these terms have been used in Sections 1.2 and 1.3 of this chapter, I go on to make the case that worth applies broadly to all aspects of reality and reality as a whole in part four and in part five link this to my understanding of the nature and purpose of reality and key related streams of holistic thought.

The business end of this chapter is its sixth part where I work through the practical implications of this broader, holistically grounded, understanding of worth for constraint of medical use of the human being. I suggest that in order to protect rather than degrade human worth medical use must be:

(1) productive in terms of the ends it pursues *and*
(2) compatible with various norms in terms of the means by which it pursues them, specifically the four norms of

   (a) protecting all members of the species as of equal inherent value;
   (b) protecting human nature and form;
   (c) protecting the species and its members from instrumentalization; and
   (d) appropriately accounting for relevant non-absolute principles, such as autonomy and beneficence.

1.2. The concept of personhood and its relationship to worth

Palazzani posits that in Western culture at least, the word person suggests on a common sense level 'a subject worthy of respect and protection' (2008, 89). It is thus not surprising that the question of who is a person is met by a wide range of distinct answers both philosophically and legally. However, most of these answers can be categorized into whether they reflect what is known as 'personalism,' its diametric opposite 'personism' or a relational approach to personhood.

Personalism refers to those approaches which only describe and use such things as the characteristics, potentials and abilities of human beings to indicate the notion that human beings individually and as a species have inherent worth or value and evoke the importance of protecting that. It has especially been used in this fashion by philosophies influenced by Christianity and metaphysics (Palazzani, 2008, 89). Bioethius thus talked of *rationalis naturae individua substantia* (*Liber de Persona et Duabus Naturis* chapter 3) which Aquinas reformulated as *individuo subsistens in rationali natura*, meaning that the person is simply the individual substance of rational nature (*Summa Theologiae* I, q. 29, a. 3). As Palazzani points out 'individual substance is the distinct subject which exists in itself, not inherent to anything else, or which belongs only to itself' (2008, 95).

Personism refers to those approaches which run directly counter to personalism by treating personhood as particular to those within (and potentially also beyond) the species who meet whatever are deemed to be the qualifying criteria of the particular version of it being espoused. Over the course of history one finds many instances of wholly arbitrary factors such as colour, ethnicity, sex and social class being used as qualifying criteria and thereby to ground distinctions in quality of social and legal treatment. Whilst that is now widely decried, there are various other more subtle forms of discrimination that continue to be entertained in respectable discourse and sometimes even in the construction of law. In particular, a significant amount of discourse resorts to functionalism – which is essentially the idea that human beings can be valued relative to their possession of certain functional abilities. The most obvious manifestations of this are the fact that the dead and the not yet born are not treated as full persons in many jurisdictions – and are subject to differences in treatment that generally extend to treating them as mere means in the context of medical use. However, it also frequently takes the form of a much more broadly disenfranchising emphasis on agency. Downie and Telfer, for example, argue that to qualify as being a person one must have a rational will which they suggest entails having,

> in the first place, the ability to choose for oneself, and, more extensively, to formulate purposes, plans and policies of one's own. A second and closely connected element is the ability to carry out decisions, plans or policies without undue reliance on the help of others.
>
> (1969, 20)

Variations on this theme include the requirement for: self-consciousness (Tooley, 1983; McMahon, 2002); self-consciousness and rationality (Buchanan, 2009, 324); being more specifically able to value one's own existence (Harris, 1985); and being able to manifest intellectual activity, freedom, moral sense and self-determination (Engelhardt, 1996). Whilst most functionalist thinkers specify a threshold of capacities that must be met, the capacities in question tend to be obtained by degrees and one could therefore be a personist and protect rights relative to that. Alan Gewirth's work, for example, fits into this category.

These approaches are invariably used to question whether even born living members of the human community should be classed as persons where they have significantly reduced capacities. However, one can argue that they are even more exclusive than this when taken to their logical conclusion. Measured against rationality or agency, for example, we are all non-persons not only prior to birth but for some time after it, at which point if we do not have severe impairment of our mental abilities we will qualify as persons but only up to the point of death or some other kind of prior permanent loss of mental capacity and arguably not during periods of temporary loss of our abilities such as due to illness, intoxication, unconsciousness and even sleep.

It may also be argued that, whilst judging between individuals is necessary in many contexts, in only a limited number of these will doing so principally or solely by reference to abilities be appropriate. In most, other factors will be at least as if not more important.

If, for example, one is choosing a friend wisely one ought perhaps to centre heavily on character traits (Vehmas, 1999, 114). Furthermore, whilst it may be necessary to judge between individuals in specific circumstances that does not make it necessary, relevant or even possible to judge between them in a more overall worth based sense. Indeed, to the contrary, it may be argued that with the very nature of our worth being inherent it is impossible to do so.

Functionalist approaches and personism as a whole can also be criticized for lack of ethical and political perspective. Expressing and building our humanity by communing with and showing compassion for ourselves and others is of principal importance for our survival, let alone ability to thrive as essentially spiritual beings. The intellectual and emotional process of categorizing our differences can be a useful one when contained in a framework of recognition of our common humanity. However, it is a dangerous process when used to promote inequalities in treatment. To see this one only needs to reference one of the more extreme manifestations of such promotion – namely eugenics which in certain forms has had at its core the 'evaluative logic' that '(s)ome human life was of more value – to the state, the nation, the race, future generations – than other human life' (Bashford & Levine, 2010, 3–4). Capacity based approaches to personhood clearly cannot be viewed as a harmless backwater of academic and social discourse but reflect and can be reflected in real world shortfalls in the universality of protection of dignity and rights.

As already noted there is also a third broad type of approach to personhood which is to treat it as being relational in character. The basic premise posited for doing so is that the moral value or worth of humans 'is not found in their individual capabilities or in their membership of the human species, but rather in their relationships' (Foster & Herring, 2017, 35). Central to this approach is a recognition of the fact of human vulnerability and the importance of an ethics of care. Nonetheless, it is possible to have these without reducing our notion of ourselves to our relationships. The starting point for doing so that is advocated later in this chapter is to consider ourselves as being both distinct aspects of reality and fundamentally interconnected with other aspects of reality and reality as a whole. Our worth in this view cannot be reduced to the value of our relationships but stems from the simple fact that we are an aspect of reality. The nature and purpose of individual aspects of reality and of the whole is what provides clues as to how we might ideally live and as part of that shapes how we might handle medical use.

1.3. Dignity as a proxy for worth

The concept of dignity was first used in Western discourse in the ancient Graeco-Roman world. The word *dignitas* sums up the variant meanings that were given to it with respect to human beings. It means wholeness or nobility in Latin, but as Milbank notes (2013, 193), it also etymologically close to *decus* which as well as meaning 'ornament' means 'honourable reward,' *decorum* meaning socially acceptable ethical style and ultimately the Greek *dokein* meaning 'to show,' *doxa* meaning 'shining manifestation' and *axia*, meaning 'fundamental worth' or 'first principle' (hence the term 'axiom'). *Decus*, *decorum* and even *dokein* reflect the emphasis placed on recognition and standing in the community. The standing of an individual might be influenced by factors particular to them such as their merit, inherited status, connections, ability to use patronage and ability to instil fear in others. However, it was also linked to the idea that human beings as a whole should be able to hold their heads up in the company of others and be properly acknowledged by them.

This civic dignity entailed, 'having one's claim recognized by others, having their respect, having some measure of control over one's life, having a say in decisions and having responsibility for one's choices' (Ober, 2014, 53). At the state level a good example of it in action was the city-state of Athens following the Athenian Revolution of 508 BC. Through that revolution, the community had 'took for itself, jointly and severally, the high standing once reserved for the few members of an elite of birth and wealth' (Ober, 2014, 58). It was by no means a perfect democracy in as much as the liberties and protections afforded under it did not result in the abolition of slavery or wholly extend to women. However, its political and legal culture did bring and reflect an increased emphasis on liberty and equality, helping to maintain conditions in which citizens were able to go about their public and private business without threats to their sense of self and reputation (Demosthenes, 1990). These ideals embodied the notion that human beings as a whole have *axia* – a worth that may be described as fundamental and inherent or intrinsic. This was later explicitly advanced in the third century BC within the Stoic tradition and then in the first century BC via Cicero's use of the term *dignitas humana* to describe what he saw as the special place of human beings in the universal order (Cicero, 1913, 6, 106).

Specific use of an inherent worth based account of dignity then became more prominent and explicit within Western formal thought in the Middle Ages.

Medieval attempts to interpret human worth relied on biblical sources and works from antiquity along with their interpretations in Arabic and Jewish texts (Imbach, 2014, 64). The former centred on the notion that man was made in the Image of God (Genesis 1:26 and Wisdom 2:23) whilst the latter used the Aristotelian notion of human beings as rational, social and political animals. Many leading thinkers of the time understood the former in a way that accounted for the latter. For example, Hugo of Saint Victor (c.1096–1141) emphasized *imago* according to *ratio* (*Da sacramentis* I, 6, 2); Bernard of Clairvaux (1090–1153) argued that free will was the essential quality designating humans as being in the image of God *(De gratia et libero arbitrio*, IX, 28); *and* Dante Alighieri (1265–1321) later described it as the greatest gift of the divine goodness (*Paradiso* V, 19–24) being the basis for responsibility and morality (*Purgatorio* XVIII, 64–66). However, one can contend that it is not freedom to choose per se that makes human life special but the fact that we have a teleology that we can link its exercise to. I might, for example, have the freedom to get drunk every night, but I also have the freedom to get drunk less often or not at all – and the latter options might, at least in the long term, better fit the purpose for which I exist. One of the themes that has cropped up repeatedly over the course of known written human history is the idea that one might thereby help to cultivate a good life by the pursuit of knowledge. In theology this is often linked to one's knowledge of God. For example, some early Christians such as St Augustine (354–430) saw pursuit of the former as the means to achieve the latter (*Soliloquai* II, 1, 1). This theme was taken up during the Middle Ages by Anselm of Canterbury (1033–1109, *Monologian* v. 66) and by Thomas Aquinas who suggested that the rational nature of human beings enabled them to imitate God through self-knowledge and self-love (*Summa Theologiae* I, q. 93, Art. 4) and provided them with a basis for their dignity (*De veritate*, q.25, Art. 6 ad 2).

The emphasis on rationality in such writings should generally not be taken to refer to the isolated use of discursive reason, which is analytic and reductionist in character, but the use of this in service of holistic and integrative reasoning which was known in ancient Greece as *intellectus.* According to this understanding the highest form of dignity, as later stressed by Giovanni Pico della Mirandola (1486) in his *Oration on the Dignity of Man* (trans Borghesis, 2012), consists of dedicating one's life to what essentially amounts to a co-creative at-one-ness with God. This is ultimately a state of being rather than simply a state of thinking – indeed, seeing the world purely through the lens of discursive reason can actually detract from it. Following on from the Socratic tradition, many medieval thinkers emphasized the idea that reality is not fundamentally separated from oneself but rather something one can come to understand through reflection on what one is and what one is not. Some thinkers suggested that the fruits of such self-examination will be awareness of the nullity and misery of human life. For example, Lotario dei Conti di Segni, who became Pope Innocent III, contrasted the inherent worth of human beings derived from their divine nature with what he viewed as wretched actual condition. However, equally others such as William of St Thierry (1080/85–1148) and Hugo of Saint Victor (c.1110–1173) considered it to be like discovering a rich new world of depth and beauty. It is, of course, a journey that is not the same for any two of us, but from my own experience I would conclude that whilst it might feel at certain points as described by the former it is ultimately more akin to the latter.

Despite its richness, medieval discourse concerning dignity suffered from a number of problems. One of these was a tendency to ignore dignity in favour of (hoped for) general deterrence in thinking about criminal punishment, which extended to the death penalty. Another was that it did not fully embrace equality for all. Formal thought about dignity was almost exclusively male and in some ways highly misogynistic in its content. It included, for example, the 'reactualization of the Aristotelian notion of the woman as a failed man' (Imbach, 2014, 70). Equally, slavery was not properly opposed and actually defended by some leading thinkers, including Thomas Aquinas who believed that it could be deduced from natural law (*Summa Theologiae* II, q. 57, Art. 3 ad 2).

By contrast, the theme of universal equality appears to have gained ground in formal thought by the late medieval period, not least through the works of such mystics as Marguerite Porete (died 1310) and Meister Eckhart (c.1260–1328). Mieth (2014, 77–83) has noted how such writers also went on to influence work on political conflict and human rights – not least coming out of the School of Padua and later the Cracow Academy. Paulus Wladimiri (c.1370–1435), who led the Cracow Academy, emphasized that whilst sin casts a shadow on dignity it cannot annul it. In applying his thinking to the context of war, he argued that it was accordingly an evil to be avoided and reached the then quite radical conclusion that even the territorial integrity of Pagan states and areas under Pagan control should be respected along with the dignity and rights of Pagans themselves.

This thinking was later reflected in work of Luther (1483–1546) and then a number of Dominican theologians working at the University of Salamanca and Jesuit theologians working at the University of Coimbra. However, the Catholic Church officially maintained the position that non-believers were not fully dignity and rights bearing human beings, treating their lands as *terra nullius* (not inhabited) and therefore capable of being even violently taken and retained by Catholics under the doctrine of discovery. This set the ideological stage for Conquistadors to be able to kill millions of indigenous people across Central and Southern America, subject millions more to such practices as rape, torture and enslavement in the course of warring upon them and to subject countless more to the *encomienda* system, under which their lands were colonized and taken and they were subject to 'enforced conversion to Christianity, compulsory labour, tribute and violent treatment' (Pharo, 2014a, 95).

As the modern era began dignity continued to be a theme within Western thought. In the seventeenth and eighteenth century natural law was a significant vehicle for this. The expression 'dignity of human nature' was commonly used in the period to describe both the idea that human beings inherently had dignity whatever their behaviour and the fact that to the extent to which they lived according to their natural capacity for virtue they had moral dignity (Masferrer, 2016, 248). The former seems to hark back to the Greek notion of *axia* and

Roman concept of *dignitas humana* whilst the latter would have been familiar within the ancient Graeco-Roman world as one version of meritocratic dignity. Both also resonate with certain modern categorizations of dignity, such as Robert Andorno's concepts of intrinsic and ethical dignity (2009, 73). Influential natural law theorists emphasizing dignity in the seventeenth century included Hugo Grotius (1583–1645), who saw it as extending to the dead, arguing that this was the logical underpinning of the reverence shown to them in the context of his consideration of the custom of burial (Grotius, 1625, 1901 ed., II, XIX, II, 4), and Samuel von Pufendorf (1632–1694) who stressed the dignity of man in the context of noting that it is natural for human beings through free will to pursue not only what was simply necessary for their survival but much more such as their gifts and talents, the company of others and so on (1660, 1931 trans., 235). Meanwhile, there was a significant natural law emphasis on the inalienability of rights flowing from the dignity of human nature in the eighteenth century through the works of the likes of Thomas Gordon (1692), Henry Home (Lord Kames) (1696–1782), John Adams (1735–1826), David Hume (1711–1776) and Edmund Burke (1729–1797).

Dignity also played a significant role in social reform efforts in the late eighteenth and nineteenth century. It is reflected in the works of such important social reformers as Thomas Paine (1791) and Mary Wollstonecraft (1790, 1796). Whilst neither the US Constitution or the French Declaration des Droits de l'Homme et du Citoyen (1789) mention dignity, they reflect a fairly advanced conception and protection of human worth through the extent to which they protect civil and political rights and freedoms. The Art. 1 emphasis of the latter on all being born free and equal helped the movement to abolish slavery within the French empire gather steam in the early nineteenth century, its efforts coming to fruition via the 1848 Proclamation of the Abolition of Slavery in the French Colonies, which explicitly draws on dignity (Barker, 2011, 27). Inherent worth can also be seen as the implicit basis of the Declaration's civic and meritocratic emphasis on all citizens, 'being equally admissible to all public dignities, places and employments, according to their capacity and without distinction other than that of their virtues and of their talents.' Liberal and socialist movements took built on this emphasis on formal equality by stressing the need to address the problems caused by substantive inequality.

Marxism was particularly powerful in this regard. Marx did not explicitly base his ideas on dignity, but it is implicitly reflected in his view that the structuring of society should meet the needs of the entire human being – i.e. 'his corporeality and his social and intellectual needs and capabilities' – should be catered to (Lohmann, 2014, 131) and also his related critique of capitalism as responsible for the destitution, deprivation, exploitation, alienation and disenfranchisement of the working classes.

By the late nineteenth century, the Catholic Church was starting to drop its defence of the Ancien Régime and itself start to address issues of equality. Account of human rights and economic justice have gone on to become central to Catholic social teaching (Hollenbach, 2014, 252) albeit that the Church was still fighting against such forms of egalitarianism as liberalism, socialism, democracy and the emancipation of women until well into this century (Rosen, 2012, 51) and even today it struggles with some of the implications of the latter and with LGBT rights.

From the middle of the twentieth century further emphasis on inherent worth was reflected in the sprouting of human rights instruments and frameworks. Most of these instruments explicitly mention dignity, but one finds differences in the extent to which they rely on it. For example, whilst both the Universal Declaration of Human Rights (UDHR) and the International Covenant on Civil and Political Rights (ICCPR) stress in their preambles that they were developed partly in recognition of the 'inherent dignity' of all members of the human family, it has been contended that dignity was merely inserted into the UDHR as a placeholder (McCrudden, 2013, 2) perhaps reflecting the fact that its architects seemingly struggled to agree on a common philosophical basis for rights (Morsink, 1999, 284–302). By contrast, the preamble of the ICCPR explicitly states that rights are derived from dignity. Health based international human rights instruments such as the Universal Declaration on the Human Genome and Human Rights (UDHGHR), the Universal Declaration on Bioethics and Human Rights (UDBHR) and the Convention on Human Rights and Biomedicine (CHRB) seemingly occupy a space between these two positions. It elaborates on the importance of dignity in more detail than the UDHR and clearly using it as more than a placeholder but not treating it as the singular fountainhead of their rights and protective provisions. Perhaps the most important is the CHRB which not only contains reference to dignity in its long title, preamble and first article but treats it as 'the essential value to be upheld' albeit hedging that assertion somewhat by starting that it is 'the basis of *most* of the values emphasised in the Convention' (Explanatory Report, para. 9).

Many jurisdictions have also included dignity within revisions of their domestic constitutional arrangements over the course of the last century. The Weimar Republic stressed it in the context of economic life in Art. 151 of its 1919 constitutional framework and following the end of World War II it went on to be emphasized in new constitutional frameworks for Spain (Art. 1 of Basic Law of 1945); Italy (Art's 3, 27 and 41 of Constitution of 1948); and the newly constituted Federal Republic of Germany (Art. 1 of Basic Law of 1949). Between 1945 and 2012 the total number of constitutions mentioning rose from 5 to 132 – the latter figure representing 84% of the world's 193 sovereign countries that are members of the United Nations. In a number of recently reformed constitutions, it is not just being mentioned in a generalized fashion but used in a variety of specific contexts. For example, the Swiss constitution applies it various biomedical contexts (Art. 119), and the Cape Verde constitution applies it to conditions of organization of work (Art. 63(1)). Dignity has also increasingly found its way into various more specific areas of legal thought and practice such as biolaw (see Chapter 3), legal psychology (Shultziner & Rabinovici, 2012); socio-legal theory (Kamir, 2006); and international criminal justice (Benton, 2012).

1.4. A broader approach to worth

Beyond largely Western debates about personhood and dignity, there are inevitably other traditions that have a history of worth related discourse and practice. A range of ancient indigenous civilizations in the Americas sought to cultivate values relating to it (see further Pharo, 2014b, 149). Some of these civilizations managed to do so in a wider-ranging fashion, partly by going beyond timocracy. For example, the Mixtecs of southern Mexico, emphasized mutual respect, freedom and protection and such values as solidarity, hospitality and communality (Pharo, 2014b, 152). Further north, several hundred years before the European invasion, five nations known collectively as the Haudenosaunee (or the Iroquois to outsiders) formed what seems to have been the world's first supra-national constitution, known as the Great Binding Law of Peace. This emphasizes sovereignty, peace, democracy, consensus and such principles as universal equality, reason, honesty and justice.

The hypothesis that emphasis on human worth is universal rather than simply a Western theme is also bolstered by the fact that there are 'terms to express similar conceptions' to dignity 'in most languages, including ancient ones' (Rosen, 2012, 11). Some Western thinkers have used non-Western reflections on worth to expand the horizons of their views about dignity. Metz, for example, has articulated a theory of dignity grounded on sub-Saharan values, specifically Ubuntu, which he suggests 'could serve as a genuine rival to the influential Kantian conception that currently dominates most intellectual reflection on the topic' (2014, 310). Echoing the Kantian notion that one should seek self-improvement, sub-Saharan thought emphasizes that one is able to develop ones attributes to become more fully a person or human (though this is not to conversely say that it views one as less of a person or human in the personism sense if one fails to do that). However, in contrast to most Western thought, it has a particularly strong communitarian emphasis on that process being brought about through relationship with other persons. As Metz puts it, '(a) large swathe of sub-Saharan thinking about ethics is summed up with phrases usually translated as a "person is a person through other persons" or "I am because we are."' (2014, 310). Aside from God, the we in this context that one should commune with is particularly ancestors and living human beings though the vital energy or life force that comes from God is also recognized in descending degrees as with animals, plants and other material forms such as rocks. Since this is essentially a felt way of living, it is easy to agree with Takyi-Amoako and Assié-Lumumba (2017, 63) that the notion of acting without humanity or dignity are unthinkable in the Ubuntu tradition.

Western and non-Western thought about worth are evidently alike in bringing us back to the question of its scope of application. Is it confined to some human beings or relevant to all human beings or relevant also beyond human beings? Graf-Vitzhum argues that religious thought generally support the former view, positing that not only Christians but also '(o)ther religions (and legal orders) similarly do not consider the animal as such to be the image of God and worthy of respect' (2004, 102). However, this is an imprecise assessment on a number of levels.

Firstly, it is only within the Judaeo-Christian tradition that human beings are said to be made in the image of God and even in this tradition, there is a tendency to also emphasize respect for non-human aspects of reality. Literal analysis of some translations of Genesis 1:26–28 suggest we might do as we please with animals and the environment, but there are important theologians, such as Martin Luther, who treat even these same parts of its first chapter as merely making humans first and foremost amongst their fellow creatures and resulting in them owing a duty of stewardship toward them (Bayer, 2014, 101). There are also numerous other parts of both (what Christians call) Old and New Testaments which support the view that human beings should respect animals and the environment. To give just two examples, Proverbs 12:10 states that '(t)he righteous one knows [the needs of] his animal's soul' and in Luke 12:27–28 Jesus is said to have emphasized respect for the majesty of the natural world by imploring us to '(c)onsider how the lilies grow. They do not labour or spin. Yet I tell you, not even Solomon in all his splendour was dressed like one of these.' A number of strands of Christianity have developed an increased sense of reverence for the idea that animals and the environment should be valued. For example, it is now a central theme of Catholic social teaching, partly because successive Popes from Paul VI (1971, 416–417) through to Pope Francis (2015) have emphasized the importance of nature and threats posed by its degradation Similar sentiments have also been expressed by the first amongst equals leader of the Eastern Orthodox Church, Patriarch Bartholomew 1 (2012).

Secondly, other traditions also embrace the inherent worth of non-human aspects of reality. The Islamic tradition similarly combines the assertion that human beings are 'the creation that God loves most' (Sheikholeslami, 2004, 11) with an emphasis on care and respect for animals and the environment. The great Eastern traditions also contain this dual emphasis. They consider being a human to be a special opportunity to grow toward a state of at-one-ment with the uncreated creator but also stress the need to tread lightly on the earth and show respect for all living things. Buddhism, for example, is said to place 'the greatest value on man, who alone of all beings can achieve the supreme state of Buddhahood' (Kyi, 1991, 174) but equally emphasizes reverence for nature and all living things.

There is also a more general growing secular emphasis on protecting the environment and animals, with activist and broader public leanings in these directions nudging state action. Sometimes such action extends to constitutional protection. For example, Art. 73 of the Federal Constitution of the Swiss Confederation states that '(t)he Confederation and the Cantons shall endeavour to achieve a balanced and sustainable relationship between nature and its capacity to renew itself and the demands placed on it by the population,' and Art. 120 protects human beings and their environment against the misuse of gene technology and requires the Confederation to legislate on the use of reproductive and genetic material from animals, plants and other organisms, taking into account the dignity of living beings as well as the safety of human beings, animals and the environment and to protect the genetic diversity of animal and plant species. The 2008 constitution of the Republic of Ecuador devotes a whole chapter to the rights of Pacha Mama (mother nature) which is seen as having: The 'right to integral respect for its existence and for the maintenance and regeneration of its life cycles, structure, functions and evolutionary processes' (Art. 71) and the right to be restored (Art. 72). It provides some direct protection against threats to the genetic constitution of existing forms by forbidding '(t)he introduction of organisms and organic and inorganic material that might definitively alter the nation's genetic assets' (Art. 73). Similarly, in 2012 Bolivia passed the Framework Law of Mother Earth and Holistic Development for Living Well which aims, amongst other things, to link the three issues and to guarantee the ability of the components and life systems of Mother Earth to regenerate. Both are capable of being linked to the broader idea that we must protect the nature and telos of reality as a whole, one of the benefits of this being that we would avert the now existential threat posed by conceptions of human interests that are not only at odds with the environment but also as a result ultimately self-defeating.

1.5. Linking a broad approach to worth with the nature and telos of reality

How we think about worth is ultimately grounded in how we think about the nature and telos of reality as a whole. Like many, I have a theistic perspective. What I call (for the purposes of neutrality) the uncreated creator exists and creates out of itself not at some physical level but rather at a spiritual one. What we can perceive with our senses is merely secondary physical aspects of a more primary spiritual reality, and as part of that, as human beings, we may most accurately regard ourselves as being souls who are experiencing a physical dimension as opposed to physical creations out of which a consciousness or soul in some unexplained and mysterious way emerges. In keeping with various holistic traditions of thought, I regard reality as a whole and all its constituent aspects to be conditioned by an entelechy of growth and to have an interconnection that means growth in one encourages growth in others and the whole and vice versa. The process of growth beings with it the freedom to make both choices that harness and dampen the natural drive toward growth. However, ultimately, whilst the sense of being free to choose is vital, it is the extent to which one is willing and able to exercise that freedom consistently with one's growth that is critical to enhancing one's freedom in a deeper and more meaningful sense. To the extent that we live antagonistically to our own growth, we undermine our own sense of confidence which is linked to the uncovering, development and expression of our self.

How is this spiritually oriented form of holistic thinking situated? Anecdotally, it seems to sit well with esoteric mystery traditions of spiritual knowledge the roots of which predate any of the major world religions. Via that and otherwise has certain resonances those major world religions – most obviously and extensively the great Eastern religions of Sikhism, Buddhism and Hinduism but also in a more partial way in the Abrahamic traditions. It also has resonances with ancient Greek thought, in the paradigmatic shift in thinking that has taken place within the hard sciences over the last hundred years or so and has been complemented by pockets of more humanistic, transpersonal and other holistic forms of thought in the soft sciences. It is to these resonances that I know turn.

Formal ancient Greek scholarship was heavily influenced by thinkers like Pythagoras, Heraclitus and Empedocles who themselves have been said to draw on mystery traditions (Graham, 1986, 3). Hall more specifically describes the links some of these scholars had with Asian masters. For example, he notes that Pythagoras was known as 'Yavancharya, the Ionian master' who 'travelled to India and was initiated into the rites of the Brahmans in the cave temples at Ellora and Elephanta' and had a number of Hindus come to study with him 'after he had established his school at Crotona' (Hall, 1939, 49). However, formal systems of Greek thought had started to lose some of this connection by the classical period and thus came to be spread, particularly via the Romans, to Western civilization in what was arguably a somewhat debased form.

Graham argues that a shift in central usage of a number of key concepts was one dimension of this debasement (1986, 4). For example, the term 'measure' shifted from referring to having 'the measure' of the essential nature of an aspect of reality something through understanding the ratio of its inner proportions to each other and the whole (the Golden Mean or Section or what was later referred to by Kepler as the Divine Proportion) to denoting 'mainly a process of comparison with some arbitrary external standard' (Graham, 1986, 4). In this form,

> it was handed down by the Greeks to Western civilisation via the Romans who thereby inherited the notion that knowledge or *scientia* – the Latin word from which the term science was derived – means objective fact, and that such facts constitute the only valid knowledge of the world.
>
> (Graham, 1986, 4)

Correspondingly, discursive reason, which is analytic and reductionist in character, moved increasingly from being used in the service of holistic and integrative reasoning, known as *intellectus*, to being treated as the end of reasoning in its own right (Happold, 1970).

Theories of reality which undermined the role of the spiritual started to take greater hold. Perhaps most notably, atomic theory was developed by Leucippus and Democritus and re-emerged again in the seventeenth century as a foundation of the so-called seventeenth century scientific revolution, including Newton's three laws of motion and his essentially mechanistic model of the solar system. This continues to exert a significant impact on thought today. Its central tenet is that the (then thought) smallest physical unit in the universe – labelled the atom – displays no purpose and simply moves through empty space (the void) combining with other units to form larger objects and/or changing direction following collisions.

Descartes was the seventeenth century theorist who is perhaps most famous for applying this mechanistic vision to the human body (1647). He saw body and soul as linked but not in the way of a 'living body experience which the French refer to as "la chair" and the Germans as der Leib' but rather as something 'reduced to a mechanism that the soul could inspect' (Illich, 1976, 156). Many have gone even further with atomic theory to postulate that there is no soul such that what has been created (including consciousness) can be reduced to an outcome of physical processes, and even in some cases to build on this by treating even the process of creation as also reducible to an outcome of physical processes set in motion by an uncreated creator. Equally many have been consciously or unconsciously influenced by it to the end of concluding that there is no uncreated creator. In sciences related to health, the spiritually dilutive impact of atomic theory took the specific form of providing a basis for the development of a biomedical orthodoxy which:

> generally views diseases as a biochemical phenomenon that can be classified into diagnostic categories through technological methods and treated, where possible, according to standardized, objectively validated mechanisms. . . . In keeping with Newtonian physics and Cartesian dualism, the biomedical model views the body as a physical system, objectively analyzable in terms of its mechanical parts. Mechanism (the 'body as machine' metaphor) and reductionism (the reduction of illness to a set of physical symptoms) dominate biomedical thinking. Disease is an outside invader that preys upon a particular part of the body; treatment attacks the invader.
>
> (Cohen, 1998, 2)

Psychiatry and medicine have continued to reflect this model – focusing inordinately on the suppression of the outward manifestations of problems, rather than on using these as clues to help understand and, where appropriate, address the causes of these problems. Psychology meanwhile is still influenced by limited perspectives of human capacity, including those of Freud and Skinner, the respective 'fathers' of its psychodynamic and behaviourist schools. Freud concluded that humans are in the grip of a conflict between elements of self and at best can achieve an uneasy and limited homeostasis in which destructive aspects of the self are contained. Skinner even more dismissively treated human beings as purely determined and good for nothing except for behavioural control as summed up in his famous phrase 'to man qua man we readily say good riddance' (1973, 176).

However, even at the time it was being espoused, this picture of scientific dismissal of human potential and the free will that goes with it was clearly flawed on numerous levels. Bertrand Russell poetically and disturbingly described the situation as one within which science was presenting the idea that

> man is the product of his causes which had no prevision of the end they were achieving; that his origin, his growth, his hopes and fears, his loves and his beliefs, are but the outcome of accidental collocations of atoms; that no fire, no heroism, no intensity of thought and feeling, can preserve the individual beyond the grave; that all the labours of the ages, all the devotion, all the inspiration, all the noonday brightness of human genius, are destined to

> extinction in the vast death of the solar system, and that the whole temple of Man's achievement must inevitably be buried beneath the debris of a universe in ruins.
>
> (1957, 45)

The presenters of this world ask us to trust their insight that its determined nature is an objective fact when the very fact of it being determined would preclude the possibility of their insight. What are we actually supposed to make of their thought then? At best they are engaged in misguided self-contradiction, and at worst they are placing themselves in an arrogantly privileged position of being in some unexplained way seers in a world that is otherwise composed of the spiritually blind. In any event it is a good rule of practical thumb never to trust the shaping of our world to those who don't themselves trust in our potentials. They will treat as fit only to be controlled. Foucault described this social control as *biopower* (1975), noting how norms of control over the body which had been developed in medicine and other human sciences have come to pervade into all types of relationship, including that with the self.

Foucault doubted the validity of applying the natural sciences to the human sciences (1998), but it is important to stress that even if that was a valid thing to do it had already long ceased to point toward the need to apply a mechanistic paradigm of thought. Sir J. J. Thompson's discovery of the electron in 1896 debunked atomism. In 1900 Max Planck debunked materialism as a whole by demonstrating that matter is essentially an aspect of energy. Further developments supporting a non-classical understanding followed. For example, the idea that a part of reality could at one point exist in a physical and non-physical way and yet at another time exist in a purely non-physical way gained credence from 1920s experimental observations which appeared to show electrons disappearing from one orbit and re-appearing in another. Heisenberg's uncertainty principle that one cannot separate out an observing aspect of reality from an observed one underpinned a move toward rejection of the classical vision that distinct parts of reality lacked any fundamental interconnection. As Bohm put it:

> even if one supposes that the physically significant variables actually existed with sharply defined values (as is demanded in classical mechanics) then we could never measure any of them simultaneously, for the interaction between the observing apparatus and what is observed always involves an exchange of one or more indivisible and uncontrollably fluctuating quanta.
>
> (1980, 89)

As Bohm noted, this was often treated as meaning 'nothing more than an expression of the fact that there is a minimum unpredictable and uncontrollable disturbance in every measurement process' (1980, 91). However, Einstein, Rosen and Podolsky soon proved the falsity of this classical interpretation of the uncertainty principle using a hypothetical experiment. Bohm helpfully describes a simplified version of this experiment as involving a molecule of zero total spin consisting of two atoms of spin which is disintegrated by a method not influencing the spin of either (1980, 91–92). The total spin in this experiment remains at zero even when the atoms are flying apart meaning the atoms are clearly interacting when from a classical perspective they should not be. Bohm ruled out the conclusion that this was due to the influence of experimental apparatus by showing that it continues when the apparatus is re-oriented whilst the atoms are in flight (1980, 92).

Quantum theory is often said to support a 'spooky' phenomenon of action at a distance but arguably supports the much more radical idea that whilst quanta are distinct they are not otherwise separate, suggesting that time and space are not real on any fundamental level. Indeed, Bohm has suggested that, like a hologram, reality is such that one part of it and the whole can be accessed from within another part. He coined the term 'holomovement' to describe this (Bohm & Hiley, 1993). This is a theory which the neuroscientist Karl Pribram (1976) has used to explain how memory can be distributed throughout the brain rather than localized in one region. Pribram's work, like Bohm's, is suggestive of all the key tenets of holism – namely that the whole is more than the sum of its parts, that it is motivated by progressive purpose and that the interconnections between the whole and the parts are such that growth in the whole promotes growth in the parts and growth in a part promotes growth in other parts and the whole. Interestingly, it was long preceded by Goldstein's study of brain injured German soldiers which found that when one part of the human brain was injured other parts would tend to contribute to the taking over of its functions, a fact which Goldstein used to reach the conclusion that:

> Normal behaviour corresponds to a continual change of tension of such a kind that over and over again that state of tension is reached that enable and impels the organism to actualise itself in further activities, according to its nature. Thus, experience with patients teaches us that we have to assume only one drive, the drive of self-actualisation.
>
> (Goldstein, 1939)

Smuts had earlier stressed this idea three decades later in his book *Holism and Evolution*. He noted that organisms exhibit a drive toward becoming 'ever more complex and significant wholes' (1926, 105) their parts functioning in harmony, 'either naturally or instinctively or consciously toward a definite inner end' (1926, 107) such that they are able to unify or synthesize 'a new arrangement of old elements' (1926, 129). Indeed, Smuts considered this drive to also apply to nature as a whole. Transformation theory, developed by Nobel Prize winning chemist Ilya Prigogine, essentially makes the same point by stressing that all structures, even non-living ones, exhibit the impulse toward reforming in a higher order of complexity and sophistication (1985). As Ferguson puts it,

> The continuous movement of energy through the system results in fluctuations; if they are minor the system dampens them and they do not alter its

> structural integrity. But if the fluctuations reach a critical size they 'perturb' the system. They increase the number of novel interactions within it. They shake it up. The elements of the old pattern come into contact with each other in new ways and make new connections. The parts reorganize into a new whole. The system escapes into a higher order.
>
> (1982, 178)

Prigogine's theory also showed that growth would exponentially build upon itself rather like the effect of a snowball rolling down a mountain:

> The more complex or coherent a structure, the greater the next level of complexity. Each new level is even more integrated and connected than the one before, requiring a greater flow of energy for maintenance, and is therefore less stable. To put it another way flexibility begets flexibility. As Prigogine said, at higher levels of complexity, 'the nature of the laws of nature changes.' It has the potential to create new forms by allowing the shakeup of old forms.
>
> (Ferguson, 1982, 178)

In summary, the idea that quantum physics and other forms of hard science support a mechanistic and reductionistic account of reality has been debunked by numerous developments in these fields. They clearly support a holistic approach, albeit that there continues to be some debate over what this precisely means in practice (see, for example, Seevinck, 2004; Healey, 2009; Miller, 2016).

It is notable that holism has traditionally been embedded across cultures globally. Virtually all have made systematic efforts to harness it when dealing with health. For example, holistic medical traditions ranging from Ayurveda, Tibetan medicine, Chinese traditional medicine and traditional Native American medicine to Western herbalism, homeopathy, naturopathy and anthroposophy view their role largely in terms of bolstering the person's own restorative and actualizing tendencies. In this respect, they share a common goal with more humanistic and transpersonal schools of psychology which view human beings not as worthless or condemned to a limited state of normality but as capable of overcoming problems they encounter as they grow toward what is variously described in the relevant literature as self-actualization, self-realization, maturity, integration or wholeness. The principal architects of these forms of psychology were Abraham Maslow and Carl Rogers. Rogers's person centred approach to psychotherapy was designed to facilitate this growth through encouraging therapists to be congruent, warm and empathetic with their clients. He characterized psychological maturing as a state where a person exhibited

> trust in the directions of inner organismic processes which, with consciousness participating in a co-ordinated rather than a competitive fashion, carry one forward in a total, unified, integrative, adaptive and changing encounter with life and it's challenges.
>
> (1977, 248)

Maslow developed a hierarchy of needs to reflect the evolutionary process within humans which he saw as beginning with needs for life, for safety and security, emerging through needs for belongingness and affection, respect and self-respect, and ending up with self-actualization (1968, 3). He saw no conflict between meeting these needs and the well-being of others. Indeed, he argued that one of the central features of a self-actualized person was the contribution to the wider good that stems from their way of being. In other words, greater levels of self-actualization amongst persons leads to benefits all round.

1.6. Toward a practical approach to protecting human worth that can be used to constrain medical use

The two overarching preconditions for carrying out medical use

The aforegoing analysis has brought the chapter to a point where it is possible to argue that in order to protect rather than degrade or devalue human worth medical use must, *firstly*, be productive in terms of the ends it pursues *and, secondly*, be compatible with various norms in terms of the means by with it pursues them. There will inevitably be some contention over what those norms are, but I make the case for the following four: (1) protecting all members of the species as of equal inherent value; (2) protecting human nature and form; (3) protecting the species and its members from instrumentalization; *and* (4) appropriate account being given to any relevant non-absolute principles, such as autonomy and beneficence.

Productivity of ends

In the introduction, I defined a medical use of the human being as one that used medical means to serve its ends and stressed that those ends need not themselves be medical. Nonetheless, most medical use is for ends that can be described as in some sense related to one or more medical ends. It is therefore the case that the limits to productive medical use are profoundly shaped by the limits to productive human medicine.

The first point to make in this light is that population health is only marginally about the medical system in the first place and actually principally about the wider environment and social, economic and cultural conditions in which we live. Illich highlighted this over 40 years ago in his seminal book *Limits To Medicine:*

> For more than a century, analysis of disease trends has shown the environment is the primary determinant of the state of general health of any population. Medical geography, the history of diseases, medical anthropology, and the social history of attitudes towards illness have shown that food, water, and air, in correlation with the level of sociopolitical equality and the cultural mechanisms that make it possible to keep the population stable, play

> the decisive role in determining how health grown-ups feel and at what age adults tend to die.
>
> (Illich, 1976, 25–8)

A similar emphasis on broader conditions of society being critical to health exists within discourse concerned with the right to health. For example, Kretzer's extensive analysis of this right emphasizes the importance of justice which she relates not just to socio-political equality but also economic equality along with solidarity (2009, 155–178).

In spite of being limited in their impact on health, medical systems have become a central feature of economic life. Spending on it as a proportion of GDP has been ramped up to extraordinary levels in most richer countries. The most extreme example of this is the United States which 5% of total GDP on it in 1960 which more than doubled to 12.1% in 1980 and has continued to grow since, reaching a figure of 17.9% in 2017 (CMS, 2019). Higher spending does not guarantee better population health. Indeed, it does not even guarantee higher life expectancy – as is often illustrated via a comparative examination of the much lower spending but higher life expectancy rates achieved in Costa Rica (Rosero-Bixby & Dow, 2016).

The second point to make is that modern industrialized systems have fostered proliferation rather than minimization of unproductive, counterproductive and destructive practices. Indeed, Illich went so far as to argue that the way that they are designed inevitably ends up producing 'clinical damage that outweighs its potential benefits,' enhancing even as it obscures 'the political conditions that render society unhealthy' and tends 'to mystify and to expropriate the power of the individual to heal himself and to shape his or her environment' (1976, 16). He described this process as a nemesis (1976, 275). To describe the epidemic proportions of the disabling impact of professional control over medicine Illich coined the phrase *iatrogenesis* from the Greek words *iatros* (physician) and *genesis* (origin). In clinical terms by the time *Limits to Medicine* was published, the great killers of the previous three generations such as polio, diphtheria, tuberculosis, pneumonia and syphilis were no longer a significant threat. To some extent, this was due to political and technological transformations which are reflected in what doctors do and say and to very basic and cheap interventions, such as antibiotics. Yet it has falsely been connected with and been used as a justification for the development of medicine into a vast industrial complex of a type which has relied heavily on expensive buildings, systems and equipment and has elevated the status, power and intrusiveness of medicine to an extraordinary degree (de Solla Price, 1963). Illich noted the considerable adverse impact of not only mistaken, callous and contra-indicated technical contacts with the medical system but approved ones (1976, 41) including an ever-expanding number directed at people who are not even ill in the first place (1976, 22). Harms caused in the pursuit of cure is supplemented by those caused in the course of exploiting the patient and, increasingly, in the course of defending against reputational damage and liability. Since Illich's seminal work the process of medicalization has continued, with the role of the medical profession and social movements in driving it now increasingly supplanted rather than simply supplemented by 'biotechnology (e.g. the drug industry), consumers, and the insurance industry' (Conrad & Bergey, 2014, 31). It is regarded as such a significant issue in sociology that a whole field of research and publication is building up around it.

Modern medical systems have reached an intensity that is also counterproductive at the social level, aided first by the radical monopoly of the healthcare professions over most forms of medical practice and now increasingly via their role as gatekeepers for health provision (Conrad & Bergey, 2014). By devaluing working with the organic capacity of individuals to heal and cope in favour of 'heteronomous management,' the medical system has deprived the environment of those conditions that endow individuals, families and neighbourhoods with control over their own internal states and over their milieu (Illich, 1976, 133). This has reinforced 'a morbid society that encourages people to become consumers of curative, preventative, industrial and environmental medicine' (Illich, 1976, 42). As one of the consequences of this, the difficulties encountered by individuals tend to be treated as individual health problems rather than as also indicators of something systemic and therefore political in nature. This can be seen, for example, in the continuing treatment of two large and overlapping populations: those deemed mentally ill and those deemed unfit for work. The emphasis is almost exclusively on treatment and control of the former and 'writing off' the latter thereby masking what the swelling of the ranks of these two overlapping populations might have to reveal to us about problems with the way in which the medical system, the world of work and the social order more generally are operating.

We have seen above that there are various strands of thought which suggest the need for medical systems to work holistically and yet the orthodoxy that underpins industrialized systems of modern medicine is antagonistic to this. As Illich himself critically observed "therapies induce the organism to interact with molecules or with machines in ways for which there is no precedent in evolution" (Illich, 1976, 47).

The intrusive and pervasive character of industrialized medicine has also begun to progressively colonize culture. Traditional emphasis on autonomous healing and coping in the context of community support is being distorted and undermined. Pain, for example, has been transformed into a technical matter linked to artificially prolonged, opaque, depersonalized maintenance and devoid of its inherent rich personal meaning and the links that this has to one's culture, which invariably includes the idea that it represents a challenge soliciting a response from the individual (Illich, 1976, 140). A mechanistic shift has also occurred with respect to the nosology of disease:

> Descartes traced the coordinates for the implementation of the project. His description effectively turned the human body into clockworks and placed a new distance, not only between soul and body, but also between the patient's complaint and the physician's eye. Within this mechanized framework, pain turned into a red light and sickness into mechanical trouble. A taxonomy

> of diseases became possible. As minerals and plants could be classified, so diseases could be isolated and categorized by the doctor-taxonomist. The logical framework for a new purpose in medicine had been laid. Instead of suffering man, sickness was placed at the centre of the medical system and could be subjected to (a) operational verification by measurement, (b) clinical study and experiment, and (c) evaluation according to engineering norms.
>
> (Illich, 1976, 166)

The baton was later picked up and run with powerfully in the non-physical domain, with ever an ever increasing proportion of many populations being regarded as having mental illness. This in turn can be linked to a wider phenomenon of the medicalization of deviant behaviour which as Dingwall notes in his preface to Conrad's classic study of the medicalization of ADHD undermines the responsibility of that behaviour's agent and offers a solution that diverts attention from its possible societal generators (2006, X):

> ADHD, for example, becomes a failure of children rather than a failure of school regimes. The pharmaceutical-medical complex steps forward to offer a remedy that suppresses the troublesome behaviour and leaves the institutional contributions unexamined. The social power of prestigious adults prevents us from asking whether the ADHD epidemic results from the increasingly confined, boring and test oriented practices of contemporary schooling creating inhumane environments for children.

Sickness of one kind of another has in effect been defined in such a way as to be pervasive with the industrialized medical system expanding rapidly to address it firstly through the development of hospitals and today via a trend toward diffusion of clinical measurement and control through society as a whole.

Finally, the role of death in culture has been progressively colonized by medicine. Illich traces six stages in this regard, each with its own iconographic expression:

> (1) the fifteenth-century 'dance of the dead'; (2) the Renaissance dance at the bidding of the skeleton man, the so-called 'Dance of Death'; (3) the bedroom scene of the ageing lecher under the Ancien Régime; (4) the nineteenth-century doctor in his struggle against the roaming phantoms of consumption and pestilence; (5) the mid-twentieth-century doctor who steps between the patient and his death; and (6) death under intensive hospital care.
>
> (Illich, 1977, 181)

The medicalization of death has run counter to its dignity, weakened the extent to which it provides an opportunity for group solidarity enhancing social ritual and reduced the status of the dead from that of persons to closer to that of an object.

A third problem is that modern medical systems continue to organize themselves around a medical orthodoxy that is mechanistic and thus counterproductive by nature because it uses a model of 'health and illness, birth and death' that is 'built around an outmoded conceptual model of how the universe behaves' which in any case was also 'fundamentally flawed from the beginning' (Dossey, 1982, XII). Rather than gravitate towards holism, the medical establishment has remained largely hostile to it. Cohen's seminal work on complementary and alternative medicine provides several powerful historical examples of such hostility in the US context including the fact that the American Medical Association was principally set up to attack homeopaths and went on to attack osteopathy and chiropractic as cultist (1998, 17–21). These attacks were made more effective by their monopolistic licencing of medical practice as a whole, which left those who dared to attempt to help the health of others without possessing a licence open to prosecution, the threat of which proved to be very real (Cohen, 1998, 22). Similarly, controls on food and drugs interfered with the capacity of people to administer to their own health (Cohen, 1998, 23). A good European example of such hostility is the way in which herbal medicines have been sucked into the orbit of EU medicine safety and quality rules, threatening the livelihood of small suppliers notwithstanding the fact that traditional herbs and mixtures are exempted from the safety rules (Commission Decision 2008/911/EC).

The means by which the prospectively productive end is pursued must be consistent with norms that flow from protecting worth

In the previous section, I posited the view that human beings not only have free will but the capacity to exercise that in a manner that is consistent with or thwarting of growth and through it their freedom in a more ultimate and meaningful sense. Directing oneself at the former rather than the latter can be said to be the ultimate mark of respect for the purpose and hence worth of both oneself, other aspects of reality and reality as a whole because growth in one parts begets growth in others and the whole. However, the important notion here is choice. Choosing to exercise freedom toward the end of growth is responsible and virtuous whereas following someone else's blueprint of a good life under duress or conditioning is not. Furthermore, it is counterproductive to attempt to impose or condition human beings towards a blueprint. Three of the most important reasons for this are:

(1) The route toward growth for each individual is profoundly individual and is brought about partly through 'artfully playing' with one's freedom;
(2) Human beings as freedom-loving creatures will ultimately tend to unconsciously or consciously rebel against efforts to direct them anyway, particularly where these lack an exceptional justification; *and*
(3) The act of exercising control over others in circumstances and to a degree that is not strictly justifiable reflects as well corruptively fosters a disrespectful mindset of excessive paternalism and relatedly authoritarianism (even

totalitarianism at the state level) that in turn weakens relevant input into decision-making and thereby its quality and ultimately also runs counter to the broader cultivation of skills of decision-making and leadership that are crucial to societal development as a whole.

Support for the notion that freedom is central to protecting human worth is one of the reasons that not just libertarians but also those who are simply broadly speaking socially liberal have concerns about dignity based discourse. Dignity acts like a fridge door for those of an eclectic range of ideological persuasions and varied levels of commitment to freedom to magnetically affix their varied lists of desired proscriptions. At the same time, however, it is important to recognize that many campaigns for specific freedoms have drawn implicitly and in some cases explicitly on dignity. This chapter, for example, has given a flavour of how dignity has been at the root of movements away from slavery and more generally toward the broad legal recognition of civil and political rights. It is also at the centre of movements toward better protection of so-called social and economic rights. In this light, one may conclude that it is more appropriately viewed as not incompatible with appropriate protection of freedom per se but rather as a field in which an overall battle and numerous specific conflicts over how and when to protect freedom are being fought out (see, for example, Beyleveld & Brownsword, 2001, and Rosen, 2012).

This sense of dignity being something that is more rounded than freedom but still heavily influenced by it is captured in Kant's approach to it. Kant distinguished between an imperfect (i.e. non-enforceable) duty of virtue on the one hand and on the other a perfect (enforceable) duty to respect the dignity of our own humanity and that of others as an end rather than treat it as a mere means (Kant, 1797, 1991, 449). Wood notes that the latter so-called Formula of Humanity takes the form of a rule or commandment, what it basically asserts is 'the value of something existing, which is to be respected, esteemed or honoured in our actions' (Wood, 1999, 141). Earlier I argued that a key element of respecting, esteeming or honouring human beings is to place a strong emphasis on their freedom, and it may be argued that is effectively following in the tradition of Kant. The imperfect character of the duty of virtue suggests that whilst he was extolling the notion that each individual should live a good life he was also eschewing the idea that notions of what this was could simply be imposed. Furthermore, strongly protecting freedom can be seen as central to Kantian thought in as much as it provides the necessary context for the exercise of rationality which Kant considered to be the foundation for morality, the capacity for which Kant controversially considered to be a precondition for having dignity (2002, 42). However, Kant equally clearly recognized that protecting freedom could be a double edged sword and envisaged circumstances in which protecting it would clash with dignity. Margalit characterizes violations of dignity as involving humiliation of a specific kind that entails 'rejection from humanity' (1999, 135). It is easy to see that there are certain forms of conduct that involve such rejection notwithstanding their being consistent with freedom. Tossing a dwarf who has freely agreed to be tossed is an oft cited example because it entails effectively treating the dwarf as figure of fun based on their physical characteristics which in turn are related to disability (see further *Manuel Wackenheim v France*, Communication No 854/1999, U.N. Doc. CCPR/C/75/D/854/1999 (2002). However, there are plenty of other less stereotypical violations of dignity that do not in any obvious sense involve such rejection. Some of them do not per se even entail the involvement of others, such as self-harm and suicide. And some of them are principally about what is done to the species as a whole as distinct from any individual or group. Accordingly, I would more openly and broadly suggest that avoiding the violation of dignity or worth is about protecting against conduct that devalues or degrades at the human level via what it does to an individual, a group, the species as a whole or any combination of these. In the medical use context specifically this means not just that the ends pursued must be productive but that they must be brought about without transgressing norms imperative to the protection of dignity or worth – of which I have already suggested there are four, which I will now consider in turn.

(a) Protecting all members of the species as of equal inherent value

The theoretical position I have already advanced is that reality as a whole and all of its constituent aspects have profound worth. This is reflected in some forms of thought and even law but clearly at odds with others, including those grounded on personhood and many of those grounded on dignity. My contention is not that it is impossible for certain aspects of reality to have special worth but that the case for doing so and reflecting that in differences in protection has not been made out. As we have seen, it has typically taken the shape of the idea that human beings as a whole are special in such a way as to uniquely possess worth or dignity or alternatively that the label person and the protections that tend to derive from it should be exclusively reserved for certain classes of human being according to assigned criteria.

Problematically, it is clear that Kant's work fits into one of these two exclusionary traditions. Which, however, is open to contention. One suggestion is that he was using the fact that rational autonomy and hence morality are normative attributes of the human species as a basis for arguing that humans have dignity as a species, as opposed to seeking to make dignity the exclusive preserve of those specific human beings who are able to exhibit them. Graf-Vitzhum, for example, asserts that dignity in the Kantian view 'is independent from the intellectual or physical character of the specific human entity, from any personal achievements, from success or failure of identity formation' (2004, 95–96) such that, '(e)ven when a human being is de facto incapable of self-determination, it remains in the human species, based on its ethic-autonomous potential, protected in a permanent, indispensable core of personality' (2004, 99). However, this may be wishful thinking. Kant's work has been repeatedly challenged not just for permitting animals to be treated as mere means but also specific classes of human beings who fall short of being fully rational, such as children and incapable adults (Hill, 2014, 218).

One way to mitigate the exclusionary power of a rationality based approach is to value its capacity to be cultivated and recuperated. There is a certain logic to doing so, notwithstanding Kerstein's view that 'it is not clear that expressing disrespect for the capacity to develop rational nature amounts to expressing disrespect for rational nature itself' (2014, 225).

However, even at this point, one is still left with at the very least the dead being outside the scope of protection as well as at least some non-human aspects of reality. Emphasis on rationality is not part of the solution of how we should treat ourselves and other aspects of reality but part of the problem – it very much reflects the predominant mechanistic emphasis of thought about human beings and perhaps also the bias of theorists themselves toward the currency they live on. Failing to protect the equal moral worth of all members of the human species is clearly incompatible with worth or dignity.

(b) Protecting human nature and form

All forms of life have a particularity which is worthy of respect. As Hauskeller has shown, the case for such respect is often underpinned by reference to concepts like integrity, identity and dignity (Hauskeller, 2007). At the human level, such respect can be seen as entailing questioning, and where necessary constraining, conduct that modifies physical and/or psychological nature whether at the individual, group and/or species level. At the level of reproductive processes, such fundamental modification tends to be physical, specifically genetic. It includes activities like cloning and hybridization within and between species and also of other forms of fundamental genetic modification such as gene engineering. This domain like many others, highlights what Jonas describes as the growth in technological power generating an imperative of responsibility (Jonas, 1979).

At the level of born human beings genetic intervention is also relevant, but one could also highlight concerns about practices which cause fundamental physical alteration in other ways – such as those entailing mutilation of sexual and other body parts and those bring about fundamental psychological change via physical means, such as lobotomy, ECT and even some of the drugs used in the treatment of mental disorder.

Treating the protection of natural forms seriously would imply the need to question the justifiability of resorting to these and other similar practices. Those that relate only to the individual require a nuanced legal treatment that factors in level of impact and autonomy. For example, one might advocate that the law regulate such practices as tattooing and ear piercing rather than prohibit them but ought to prohibit the most extreme forms of psychology altering mental health interventions even when these are being agreed to by a capable adult. However, autonomy should be treated as much less significant when it comes to activities that go beyond threatening the natural form of a given individual such as those involving genetic modification of the human form. There is a strand of discourse within the scientific and philosophical community that belittles arguments about naturalness of form even when they relate to something as serious as such modification. The advocates for allowing it often use the suggestion that it has caused no-one harm as a basis on which to express disdain for their opponents. However, it may be observed that plants, animals and human embryos are now being modified in radical ways and there is simply no basis on which to be secure that such modification will not have impacts that will be harmful even in the narrow sense. Can we really be secure, for example, that the modified foodstuffs that have entered the human food chain are consistent with human health? Can we be secure that allowing those who will pass on mitochondrial disease down their line to avoid doing so by modifying the embryos they create via the introduction of genetic material from a third party, will result in healthy children who if or when they in turn reproduce will not pass an entirely new set of genetic problems down their line?

Since the answer to this question must be no, it can be said that rather than paying due respect to the precautionary principle, we are already playing a kind of genetic Russian roulette with our future in violation of Jonas's specific formulation of his imperative of responsibility: 'act so that the effects of your action are compatible with the permanence of human life on earth' (Jonas, 1979, 11). Furthermore, when we understand harm properly in its broader sense, we must also go beyond looking at specific potential health effects to assessing the impact of actual and proposed practices on important principles and values which in this context include protection of worth or dignity and certain other related norms, including identity and integrity. As Hauskeller notes, concerns about the manipulation and modification of living beings rest not only on hazards posed by doing so but also on the very idea of it (2007, 1) – in other words people wonder whether it is inherently wrong. The same point can also be made for plants. It can be made not just in secular terms but by reference to the spiritual idea that forms are the way they are because this is so designated by an uncreated creator. The Islamic tradition hits this point on the nose in the Qur'an 30:30 which states that 'there is no altering the creation of Allah.'

(c) Protecting the species and its members from instrumentalization

Instrumentalization in the human context describes a way of interacting with others which in informal language would be known as using them and in more formal Kantian language would be said to involve thingifying them by treating them as a mere means to one's ends as opposed to treating them as ends in their own right. In a bodily context we necessarily share interests in physical space, but it can be contended that to respect us as ends means interference with our other body related interests is only acceptable to such extent as is lawful and strictly necessary to proportionately account for the conflicting rights of another, others or society at large. The types of situation in which such restriction might take place include where individuals engage in/are legitimately suspected of having engaged in conduct that impacts the fundamental rights of others (e.g. paternity testing where the man being tested is legitimately suspected to be the father, taking DNA samples in certain circumstances from those alleged to have committed a crime) or have/are legitimately suspected of having a condition which threatens the fundamental rights of others (e.g. taking measures to control an individual's movement where they have/it is legitimately suspected that they may have a contagious disease).

Restrictions of these kinds are quite different in character to those which are predicated merely on meeting the demands of others – even when these demands can be linked to genuine needs. As we will see in more detail in Chapter 3, the distinction was famously pointed out in the case of *McFall v Shimp* (1978) No. 78–17711. 10 Pa D & C (3d) 90 (Ct Comm Pl, Pa) where in the Alleghany County Court, Mr Justice Flaherty distinguished between moral and legal duties to help others and specifically stressed that the idea of imposing a legal duty to give up one's freedom from bodily intrusion to meet the need of another went against respect for the individual and raised the spectre of the 'swastika and inquisition' when it reached the levels of removal of living body tissue. Some ethicists have tried to counter that Kant's vision of treating human beings as ends is compatible with a certain degree of dilution of protection of their body related interests to meet the needs of others. For example, Fabre has suggested that such use does not 'in and of itself imply that one is not regarding them as having their own projects and attachments' (2006, 113). However, to view the use of others as only inherently problematic where it completely denies the individuality of a human being is to tantamount to concluding that only slavery is inherently problematic. Kant's view that humans should not be treated as mere means was clearly intended to have much wider implications than that as is evident from the fact that it was set within an emphasis on protecting the dignity of humanity. It was clearly intended to cover exploitation in general – not merely what the most extreme forms of it.

All forms of exploitation entail thingification and in this sense commodification because they promote the use value of subjects at the cost of their inherent value. However, the term 'commodification' is also often used more selectively to describe thingification of a kind that entails the provision of money or money's worth for the use of bodies and body parts at a level that goes being legitimate defrayment or reimbursement of reasonable actual expenses to constitute a potential inducement or, in some cases, even a source of profit.

Protecting against commodification in its narrow sense is sometimes treated as being at the very core of an agenda of protecting worth or dignity. Thus, for example, one finds less reluctance to protect the most marginal classes – i.e. the dead and not yet born – from it than to protect their worth or dignity more generally. A good case in point is that whilst European law allows pre-primitive streak embryos to be instrumentalized to the end of stem cell production it does not allow those stem cells to be patented – specifically in *Oliver Brüstle v Greenpeace e V C-34/10*, the EU Directive on the legal protection of biotechnological inventions (98/44/EC) was interpreted as not extending to such patenting on the basis that to permit it would fall foul of the protection of dignity contained in Art. 1 of the European Charter of Fundamental Rights 2000. However, whilst there is now extensive action being taken against commodification in its narrow sense in the health field, there are also gaps which suggest ambivalence. Part of the problem is that modern medicine and society aspire to protect against commodification (in its narrow and broad sense) but subvert that by enabling production and exchange of commodities to extend to the human body being treated as both raw material and commodity (Decker, 2014).

(d) Appropriately accounting for relevant non-absolute principles

Arguably the most important of these principles is autonomy. There is no general consensus on what autonomy means, but in this context I am using it to refer to self-rule as implied by its Greek etymology. In this form, it is arguably slightly narrower than how the concept of liberty is typically understood which is both as freedom from intrusion and freedom of action. In any event, the general point may be made that strongly protecting autonomy serves a range of functions as I have highlighted towards the beginning of this section. Nonetheless, both ethically and legally it is normatively recognized as being less than an absolute value to protect. The extent of defence of it is at its zenith when it comes to capable adults and (at least in some jurisdictions) competent minors.[1] However, even then a distinction is made between a member of these classes simply exercising their freedom and others being involved in the facilitation of that. With regard to the former many jurisdictions will defend at least capable adults even when they are refusing the initiation or continuation of life maintaining treatment. Some will do so even when the need for such treatment was occasioned by a prior attempt to commit suicide and as such could itself be regarded as classically suicidal. For example, in *Kings College Hospital NHS Foundation Trust v C and Another* [2015] EWCOP 80, the English High Court protected C's freedom to refuse dialysis upon establishing that she had capacity (and that the intervention was not separately justifiable under mental health law as treatment for a mental disorder) despite the fact that her need for it was occasioned by the kidney and liver damage she had sustained when attempting to kill herself with paracetamol. With regard to the latter, by contrast, the protection of freedom is normally tempered by principles and rules that are designed to ensure that the conduct of others towards them is within 'acceptable parameters.'

Most jurisdictions will have both more overarching principles and rules that can be used to this end (of, for example, public order, public policy and dignity) and one that are designed to cater for specific situations (such as female genital mutilation, living donor organ transplantation etc.). Despite differences in their

1 The designations 'capable' adult and competent 'minor' and 'incapable' adult and 'incompetent' minor are used for ease at various points, though I agree with jurisdictions which eschew assigning individuals a general status in favour of functional analysis of their ability to make whatever decision or decisions are at hand.

specifics, the main overall purpose of these principles and rules and related systems for their enforcement is to place barriers in the path of those who are doing things which are significantly at odds with the worth of another or others and in this way can be called demeaning or degrading. In the use context, the added dimension is that such conduct will be exploitative. Conduct at odds with one or more of the three absolute norms would qualify as would conduct that violates this fourth norm by placing excess weight on autonomy relative to another relative principle such as beneficence.

A couple of concrete examples from the field of medical use of the human being serve to illustrate. Firstly, consider that someone wants to give up their single remaining kidney for the purpose of transplantation. It is possible that one could argue that this could be regarded as interfering with one of the absolute principles that flows from dignity – viz an argument that the impact of its loss on the body makes its removal inconsistent with the core of protecting natural form. However, this argument would clearly be controversial when one considers that other essential to life materials sometimes have to be removed for therapeutic reasons with their functions then taken over by some 'artificial means' (in this case dialysis).

An argument that would attract much more support would be that the removal of an essential to life organ is not unacceptable per se in dignity terms but becomes so where it is not justifiable therapy for the person subject to it since its maleficent effects are too great to justify it – even when the act of doing it fulfils the autonomous wishes of its potential subject. This would be irrespective of the 'class' of the potential subject – the limits of what a capable adult can do and what a competent minor can or at least should be able to do are simply greater than those for the incapable adult and incompetent minor. This can be justified because it is consistent with respect for the worth or dignity of the former classes to minimize interference with the exercise of their freedom whereas to respect the latter classes when they are prospective subjects of medical intervention or other actions one must try to interpret what is in their interests and protect these to the point of being willing to go against their wishes where this is strictly necessary. Contrast this with a situation in which a person sells one of their body parts to another person. By completing the transaction, the purchaser would be helping the seller to realize their autonomous desires but irrespective of whether or not one could an argument that the transaction was maleficent for the seller it would be off limits per se as a form of instrumentalization (of the seller by him or herself and by the purchaser

Nonetheless, it may be noted that capable adults and in some jurisdictions also competent minors are left with considerable legal freedom to be willing subjects of conduct, including medical intervention, of a kind that might 'objectively' be considered fairly inconsistent with their own interests. With other classes, by contrast, the protection of human worth is generally understood to necessitate the principle of beneficence being prioritized over autonomy. This is not to say that the values, wishes and choices of these classes are not relevant but rather to assert that their relevance is typically confined to their being factors in the assessment of what is good for them and only likely to tip the scales in cases where they are fairly finely balanced or less serious considerations are at stake.

1.7. Conclusion

Tussles between different approaches to understanding and protecting worth have long been an integral part of the tapestry of human thought. I have argued that an inclusive and powerful approach to worth logically follows on from understanding the nature and purpose of reality in a holistic fashion. I have set this argument against the background of Western tussles over what worth means and how it should be protected being played out via and shaped by discourses on dignity and personhood. However, the extensive use of these concepts as proxies for worth has not been without problems.

Personhood is suggestive of a status that is at best particular to human beings and at worst only some of their number and as such falls at the first hurdle of conveying worth as inclusive. Meanwhile, whilst there has been a growing trend towards an inclusive approach to dignity it is one held back by the continued baggage of many, not least within the religious community, who remain determined to view it as something particular to human beings – even using the prefix 'human' to this end when describing it. Nonetheless, the use of dignity as a proxy for worth has become so normative in ethical and legal discourse, particularly in the medical field, that it is legitimate to question whether it is practical to avoid drawing on it when communicating intentions to protect worth.

Conversely, whether it is of any specific value to draw on it has also been questioned. As Rosen notes, part of that practical concern about using it emanates from the viewpoint that it has 'no coherent meaning of its own' (2012, 6) and is thus a mere receptacle 'given content by a range of extraneous political, social and religious convictions' (2012, 6). In evaluating this concern, Rosen notes that variations in the use of dignity are undeniable by showing that it has in fact been used to support diametrically opposed positions on certain issues. He lists debates on assisted suicide and LGBT rights as examples (2012, 6–7), but one could equally list others such as abortion and broader control over conduct in pregnancy. However, such variations are only proof that its meaning, like that of many concepts and principles, is contested – with the aforegoing examples given by Rosen, illustrating a specific contestation over the general extent and specific circumstances in which it may be regarded as vindicating and conversely restricting protection of autonomy. As already noted, a generally liberty minded person one would naturally be concerned about its potential to be used to help justify or maintain what one would consider to be retrograde forms of restriction on autonomy. This, in turn, raises the question of whether one is better off defending a heavily autonomy infused vision of it or simply abandoning its use altogether. The answer to that question must lie partly in the determination of whether or not dignity gives us something 'extra.' On this front, it has sometimes been asserted that other principles can be left to do the work. For example, in the context of medical law and ethics, Macklin famously asserted that dignity is useless – an argument she based on her assessment that reference to it within the CHRB and other human rights instruments:

> seems to have no meaning beyond what is implied by the principle of medical ethics, respect for persons: the need to obtain voluntary, informed consent; the requirement to protect confidentiality; and the need to avoid discrimination and abusive practices.
>
> (2003, 1419–1420)

However, this assessment is inaccurate. The idea that dignity has no meaning beyond what is implied by medical ethics misses the point that medical ethics is not a principle but a field of moral thought, within which an inherent worth related account of dignity forms one of a number of contrasting and sometimes competing strands. The principle of respect for persons may also be one of these strands, but it suffers from substantive as well as linguistic associations not only with anthropomorphism but also even personism. Some who draw on it have specifically done so to support an agenda of mere means to an end medical use of some classes of human beings. For example, the philosopher John Harris has employed it in the service of such ideas as using incompetent adults in research that is inconsistent with the protection of their interests (Harris, 1997); compulsory harvesting of organs for transplantation from the dead (Harris, 2003); and even use of those deemed to be in a persistent vegetative state as an organ bank for others (Harris, 1985). Meanwhile, the role of specific principles or requirements such as voluntary and informed consent, confidentiality and protection from discrimination and abusive practice, cannot ultimately be determined against by anything but an overarching frame of reference.

Given this, it is clearly best that the frame of reference ultimately used is one that has been discussed and devised in the sunlight of open discourse. One might ultimately tactically consider an alternative frame in contexts where dignity has been heavily implicated in retrograde restriction of autonomy but even here one would need to be cautious of the fact that the alternative frame could also end up being misused. Dignity at least has a hinterland to guide its application in the form of a rich tradition of being drawn on in ethical and human rights based discourse and legal discourse more broadly.

That is not to say that it cannot be misused. Indeed, the history of bioethics and biolaw is in part a history of overly conservative and paternalistic thought and practice some of which can be said to have ultimately derived from oppressive conceptions of dignity. However, the risk of any increased emphasis on it becoming a vehicle for undermining the legitimate role of freedom within a proper understanding of respect must be balanced against the fact that, at least in most democratic countries, freedom itself has become a vehicle for undermining the legitimate role of dignity within a proper understanding of respect. There is, for example, a growing tolerance and even embrace of various practices which entail alienation of body related interests that is both significant and disproportionate to any expected counter-veiling benefits. This presents a particular significant problem in the context of medical use because it builds on existing ideological, professional, personal and commercial interests opposing constraint.

References

Andorno, R. (2009) Human dignity and human rights as a common ground for a global bioethics. *Journal of Medicine and Philosophy.* 34, 223–240.

Barker, C. (2011) *Dignity in International Human Rights Law.* International Write for Rights Movement.

Bashford, A. & Levine, P. (2010) *The Oxford Handbook of the History of Eugenics.* Oxford: Oxford University Press.

Bayer, O. (2014) Martin Luther's conception of human dignity. In Duwell, M., et al. (eds.) *The Cambridge Handbook of Human Dignity: Interdisciplinary Perspectives.* Cambridge: Cambridge University Press, pp. 101–107.

Benton, H. (2012) Human dignity at trial: Hard cases and broad concepts in international criminal law. *George Washington International Law Review.* 44, 317.

Beyleveld, D. & Brownsword, R. (2001) *Human Dignity in Bioethics and Biolaw.* Oxford: Oxford University Press.

Bohm, D. (1980) *Wholeness and the Implicate Order.* Abingdon: Routledge and Kegan Paul.

Bohm, D. & Hiley, J. (1993) *The Undivided Universe: An Ontological Interpretation of Quantum Theory.* Abingdon: Routledge.

Buchanan, A. (2009) Moral status and human enhancement. *Philosophy and Public Affairs.* 37(4), 346–381.

Cicero, M. (1913) On Duties (De Officiis): XXI (trans. Miller, W.). Cambridge, MA: Harvard University Press, Re-Issue edition (31 January), Vol. 6.

CMS. (2019) The National Health Expenditure Accounts, table 1 of highlights pdf on home page. Available at https://www.cms.gov/Research-Statistics-Data-and-Systems/Statistics-Trends-and-Reports/NationalHealthExpendData/National-HealthAccountsHistorical.html. Accessed 24.5.2019.

Cohen, M. (1998) *Complementary and Alternative Medicine: Legal Boundaries and Regulatory Perspectives.* Baltimore and London: John Hopkins University Press.

Conrad, P. & Bergey, M. (2014) The impending globalization of ADHD: Notes on the expansion and growth of a medicalized disorder. *Social Science & Medicine.* 122, 31–43.

Decker, O. (2014) *Commodified Bodies.* London: Taylor and Francis.

Demosthenes. (1990) *Demosthenes, Against Meidias: Oration 21* (ed. and trans. MacDowell, D.). Oxford and New York, NY: Clarendon Press.

Descartes, R. (1647) *La description du corps humain.* Unfinished Manuscript.

deSolla Price, D. (1963) *Little Science, Big Science.* New York, NY: Columbia University Press.

Dingwall, R. (2006) Series editor's preface. In Conrad, P. (ed.) *Identifying Hyperactive Children: The Medicalization of Deviant Behavior.* Aldershot: Ashgate.

Dossey, L. (1982) *Space Time and Medicine.* Boulder, CO: Shambala Publications Inc.

Downie, R. & Telfer, E. (1969) *Respect for Persons: A Philosophical Analysis of the Moral, Political and Religious Idea of the Supreme Worth of the Individual Person.* London: Allen and Unwin.

Engelhardt, H. (1996) *Foundations of Bioethics*. New York, NY: Oxford University Press.
Fabre, C. (2006) *Whose Body Is it Anyway? Justice and the Integrity of the Person*. Oxford: Oxford University Press.
Ferguson, M. (1982) *The Aquarian Conspiracy: Personal and Social Transformation in the 1980's*. London: Paladin (Granada).
Foster, C. & Herring, J. (2017) *Identity, Personhood and the Law*. Heidelberg: Springer.
Foucault, M. (1973) *The Birth of the Clinic: An Archaeology of Medical Perception* (trans. Sheridan-Smith, A.). New York, NY: Vintage Books.
Foucault, M. (1998) *The History of Sexuality*. Harmondsworth: Penguin History.
Goldstein, K. (1939) *The Organism*. New York, NY: American Books.
Graf Vitzthum, W. (2004) Back to Kant. In Vöneky, S. & Wolfrum, R. (eds.) *Human Dignity and Human Cloning*. Leiden; Boston, MA: Martinus Nijhoff, pp. 87–106.
Graham, H. (1986) *The Human Face of Psychology*. Milton Keynes and Philadelphia, PA: Open University Press.
Grotius, H. (1625) *The Rights of War and Peace II*. Available at http://oll.libertyfund.org/titles/553 #Grotius_0138_583_585.
Hall, M. (1939) *Reincarnation: The Cycle of Necessity*. Los Angeles, CA: Philosophers Press.
Happold, F. (1970) *Mysticism: A Study and an Anthology*. Harmondsworth: Penguin.
Harris, J. (1985) *The Value of Life*. Abingdon: Routledge.
Harris, J. (1997) The ethics of clinical research with cognitively impaired subjects. *Italian Journal of Neurological Sciences*. Supplement, 5, 9–13.
Harris, J. (2003) Organ procurement: Dead interests, living needs. *Journal of Medical Ethics*. 29, 130–134.
Hauskeller, M. (2007) *Biotechnology and the Integrity of Life*. Aldershot: Ashgate.
Healey, R. (2009) Holism in quantum mechanics. In Greenberger, D., et al. (eds.) *Compendium of Quantum Physics*. Berlin; Heidelberg: Springer, pp. 295–298.
Hill, T. (2014) Kantian perspectives on the rational basis of dignity. In Duwell, M., et al. (eds.) *The Cambridge Handbook of Human Dignity: Interdisciplinary Perspectives*. Cambridge: Cambridge University Press, pp. 215–221.
Hollenbach, D. (2014) Human dignity in Catholic thought. In Duwell, M., et al. (eds.) *The Cambridge Handbook of Human Dignity: Interdisciplinary Perspectives*. Cambridge: Cambridge University Press, pp. 250–259.
Illich, I. (1977) *Limits to Medicine: Medical Nemesis and the Expropriation of Health*. London: Mariono Boyars.
Imbach, R. (2014) Human dignity in the middle ages (twelfth to fourteenth century). In Duwell, M., et al. (eds.) *The Cambridge Handbook of Human Dignity: Interdisciplinary Perspectives*. Cambridge: Cambridge University Press, pp. 64–73.
Jonas, H. (1979) *The Imperative of Responsibility: In Search of Ethics for the Technological Age*. Chicago, IL: University of Chicago Press.
Kamir, O. (2006) Honor and dignity in the film unforgiven: Implications for sociolegal theory. *Law and Society Review*. 40, 193–233.
Kant, I. (1797/1991) *Die Metaphysik der Sitten* (trans. Gregor, M. as The Metaphysics of Morals). Cambridge: Cambridge University Press.
Kant, I. (2002) *Groundwork of the Metaphysics of Morals* (eds. Hill, T. & Zweig, I.) Oxford: Oxford University Press.

Kerstein, S. (2014) Kantian dignity: A critique. In Duwell, M., et al. (eds.) *The Cambridge Handbook of Human Dignity: Interdisciplinary Perspectives*. Cambridge: Cambridge University Press, pp. 222–229.

Kretzer, L. (2009) *Health, Solidarity and Justice: A Discourse, Theoretical Perspective*. Ph.D. Thesis. Queen Mary University, London.

Kyi, A. (1991) *Freedom from Fear*. New York, NY: Penguin.

Lohmann, B. (2014) Human dignity and socialism. In Duwell, M., et al. (eds.) *The Cambridge Handbook of Human Dignity: Interdisciplinary Perspectives*. Cambridge: Cambridge University Press, pp. 126–134.

Mackenzie, C., et al. (2014) Introduction: What is vulnerability and why does in matter for moral theory. In Mackenzie, C., et al. (eds.) *Vulnerability*. Oxford: Oxford University Press, pp. 1–29.

Macklin, R. (2003) Dignity is a useless concept. *British Medical Journal*. 327, 1419–1420.

Margalit, A. (1999) *Politik der Wuerde: Uber Achtung and Verachtung*. Frankfurt: Fischer.

Masferrer, A. (2016) Taking human dignity more humanely. In Masserer, A. & Garcia-Sánchez, E. (eds.) *Human Dignity of the Vulnerable in the Age of Rights*. Heidelberg: Springer, pp. 221–256.

Maslow, A. (1968) *Toward a Psychology of Being*. New York, NY: Van Nostrand Reinhold.

McCrudden, C. (2013) In pursuit of human dignity: An introduction to current debates. In McCrudden, C. (ed.) *Understanding Human Dignity*. Oxford: Oxford University Press, pp. 1–58.

McMahon, J. (2002) *The Ethics of Killing*. Oxford: Oxford University Press.

Metz, T. (2014) Dignity in the Ubuntu tradition. In Duwell, M., et al. (eds.) *The Cambridge Handbook of Human Dignity: Interdisciplinary Perspectives*. Cambridge: Cambridge University Press, pp. 310–318.

Mieth, D. (2014) Human dignity in late-medieval spiritual and political conflicts. In Duwell, M., et al. (eds.) *The Cambridge Handbook of Human Dignity: Interdisciplinary Perspectives*. Cambridge: Cambridge University Press, pp. 74–84.

Milbank, J. (2013) Dignity rather than rights. In McCrudden, C. (ed.) *Understanding Human Dignity*. Oxford: Oxford University Press, pp. 189–208.

Miller, E. (2016) Quantum holism. *Philosophy Compass*. 11(9), 507–514.

Morsink, J. (1999) *The Universal Declaration of Human Rights: Origins, Drafting, and Intent*. Philadelphia, PA: University of Pennsylvania Press.

Ober, J. (2014) Meritocratic and civic dignity in Greco-Roman antiquity. In Duwell, M., et al. (eds.) *The Cambridge Handbook of Human Dignity: Interdisciplinary Perspectives*. Cambridge: Cambridge University Press, pp. 53–63.

Paine, T. (1791) *Rights of Man: Part of the First*.

Palazzani, L. (2008) Person and human being in bioethics and biolaw. In Weisstub, D. & Diaz-Pintos, G. (eds.) *Autonomy and Human Rights in Healthcare: An International Perspective*. London: Springer-Verlag, pp. 89–98.

Patriarch, Bartholomew 1. (2012) *Global Responsibility and Ecological Sustainability*. Closing Remarks, Halki Summit I, Istanbul, 20 June.

Pharo, L. (2014a) The Council of Valladolid (1550–1551): A European disputation about the human dignity of indigenous peoples of the Americas. In Duwell, M., et al. (eds.) *The Cambridge Handbook of Human Dignity: Interdisciplinary Perspectives*. Cambridge: Cambridge University Press, pp. 95–100.

Pharo, L. (2014b) Human dignity and indigenous peoples of the Americas. In Duwell, M., et al. (eds.) *The Cambridge Handbook of Human Dignity: Interdisciplinary Perspectives*. Cambridge: Cambridge University Press, pp. 147–154.
Pico della Mirandola, G. (1486) Oration on the Dignity of Man (eds. Borghesi, F et al, 2016) Cambridge: CUP.
Pope, Francis. (2015) *Encyclical Laudato si'* (Praise Be to You – On Care for Our Common Home), 24 May.
Pope, Paul VI. (1971) Apostolic letter. *Octogesima Adveniens*. 14 May, 21, AAS 63.
Pribram, K. (1976) *Consciousness and the Brain*. New York, NY: Plenum.
Prigogine, I. & Stengers, I. (1985) *Order out of Chaos: Man's New Dialogue with Nature*. London: Fontana.
Rogers, C. (1977) *On Personal Power*. London: Constable & Co Limited.
Rosen, R. (2012) *Dignity: Its History and Meaning*. Cambridge, MA: President and Fellows of Harvard College.
Rosero-Bixby, L. & Dow, W. (2016) Exploring why Costa Rica outperforms the United States in life expectancy: A tale of two inequality gradients. *PNAS*. 113(5), 1130–1137.
Russell, B. (1957) *A Man's Free Worship in Mysticism and Logic*. New York, NY: Anchor Doubleday.
Seevinck, M. (2004) Holism, physical theories and quantum mechanics. *Studies in History and Philosophy of Modern Physics*. 35B, 693.
Sheikholeslami, R. (2004) The creation and dignity of man in Islam. In Vöneky, S. & Wolfrum, R. (eds.) *Human Dignity and Human Cloning*. Leiden; Boston, MA: Martinus Nijhoff, pp. 3–12.
Shultziner, D. & Rabinovici, I. (2012) Human dignity, self-worth, and humiliation: A comparative legal-psychological approach. *Psychology, Public Policy, & Law*. 18, 105–143.
Skinner, B. (1948) *Walden Two*. Indianapolis, IN: Hackett Publishing.
Skinner, B. (1973) *Beyond Freedom and Dignity*. Harmondsworth: Penguin.
Smuts, J. (1926) *Holism and Evolution*. New York: The Macmillan Company.
Takyi-Amoako, E. & Assié-Lumumba, N. (2017) *Re-Visioning Education in Africa: Ubuntu-Inspired Education for Humanity*. Heidelberg: Springer.
Tooley, M. (1983) *Abortion and Infanticide*. Oxford: Clarendon Press; Oxford University Press.
Vehmas, S. (1999) Newborn infants and the moral significance of intellectual disabilities. *Research and Practice for Persons with Severe Disabilities*. 24, 111–119.
Wollstonecraft, M. (1790) *Vindication of the Rights of Man*. London: Printed for J. Johnson.
Wollstonecraft, M. (1796) *Vindication of the Rights of Women*. London: William Pickering.
Wood, A. (1999) *Kant's Ethical Thought*. Cambridge: Cambridge University Press.

2 Alternative ethical benchmarks

2.1. Introduction

In the last chapter, I evaluated the merits of different approaches to respect and ultimately made a positive case for a holistic one that was deep and inclusive in its understanding and protection of the worth of human beings individually and as a species. I also laid out the implications of taking this approach when guiding and constraining medical use.

However, it must also be accepted that neither philosophical ethics or bioethics is wholly respect centric but rather also includes a range of alternative benchmarks for judging human conduct. The purpose of this chapter is not to identify, explain and evaluate those alternatives as a whole but to selectively focus on those that have been used relatively frequently to advocate what amount to lesser protection in the context of medical use – specifically libertarianism, justice and utilitarianism.

2.2. Libertarianism

The Cambridge Online Dictionary sums up a libertarian as 'a person who believes that people should be free to think and behave as they and should not have limits put on them' (2019). However, a more nuanced view is that it is about maximizing the scope for freedom within the inherent limits of the need to deal with the fact that when the freedoms of one individual, group or society are exercised they will at times clash with those of others.

Whilst libertarianism is often proposed as a theoretical framework based on respect, I have categorized it as an alternative because its emphasis on maximizing rather than simply heavily protecting liberty only really properly caters to our capacity for free will rather than also to the vulnerabilities that characterize our condition and the related need to generate a strong milieu of mutuality, solidarity and care. As to our vulnerabilities, it has been noted that:

> By virtue of our embodiment, human beings have bodily and material need; are exposed to physical illness, injury, disability, and death; and depend on the care of others for extended periods during our lives. As social and affective

> beings we are emotionally and psychologically vulnerable to others in myriad ways: to loss and grief; to neglect, abuse, and lack of care; to rejection, ostracism and humiliation. As socio-political beings, we are vulnerable to exploitation, manipulation, oppression, political violence, and rights abuses. And we are vulnerable to the natural environment and to the impact on the environment of our own, individual and collective, actions and technologies.
> (Mackenzie et al., 2014, 1)

Interwoven with these vulnerabilities, we can not only end up making poor decisions but also suffer relative or complete loss of capacity to make decisions at all. This latter situation is a state of affairs that a framework of ethics does not directly cater to at all where it is simply driven by an emphasis on liberty. Nonetheless, to simply reject libertarianism in totality as excessively individualistic would be unduly dismissive. Some forms of it very pointedly emphasize liberty out of a belief that protecting it is necessary for the society to flourish. Indeed, the desire that many libertarians have to see the state wither is often based on this. Ward, for example, goes as far as suggesting that, '(t)here is an inverse correlation' between the strength of society and the state and that accordingly '(i)f we want to strengthen society we must weaken the state' (1998, 20). However, in making this argument he refers, like many other anarchists, to the state being not only a narrow set of institutions holding various powers but also a mindset and way of behaving grounded in the twin pillars of dominance and submission which allows society to be pervaded by hierarchy and coercion. Consequently, the state in its overall sense cannot be destroyed simply by a revolt against it in its narrow sense but also requires different ways of thinking and behaving, centred on individuals banding together to take control over their lives, developing and sharing systems of organization that are less hierarchical, more open and more social in character.

This is an entirely laudable goal. However, it is also one that many libertarians regard as unrealistic. Nozick, for example, argued in his seminal book *Anarchy, State and Utopia* (1974) that societies that attempted to rid themselves of the state would end up relying on it in some minimal form. He concluded that having a minimal state to protect rights was a necessary evil. I take what I think is a slightly more nuanced view that actually an end to the state could be realized but that its realization cannot be forced ahead of human growth and that the force of the state will be necessary to protect rights until such time as humans can be relied on to mutually respect them without it. In the medical use context, this would include rights to be protected from degrading and devaluing conduct.

It is also notable that, whilst some libertarians, including anarchists like Ward, emphasize the social sphere and such specific values as mutuality, solidarity and care, others have supported capitalism in ways that have had a contrary influence on our relationships with each other, animals and the environment and have specifically ended up fostering commodificatory medical use. Much of the problem in this regard is that property has not been treated as a simple construct to be applied and protected only where compatible with respect for all aspects of reality and reality as a whole but as a fundamental right affording a wide-ranging set of rights of control – such as the right to keep, exchange, transform etc. – with respect to both physical and creative dimensions of reality. There are two libertarian views as to its scope in this regard.

Firstly, the extended view – which is specifically relevant to the physical dimension (of plants, animals, minerals, land etc.) – is that in certain contexts humans can acquire property rights by mere exercise of dominion. It is difficult, however, to find a credible ethical justification for this. The seventeenth century philosopher John Locke (Laslett, 1960) suggested that it is what God intended. However, if God created out of love it would seem more credible to suggest that it wanted the benefits of that creation to be enjoyed fairly by all in current and future generations as opposed to being aggressively hoarded in what amounts to little more than an extended game of 'bagsy.'

Secondly, the less extended view, vividly reflected in Ayn Rand's magnum opus novel *Atlas Shrugged* (1957) and in the works of the likes of Nozick (1974) and Machan (2006), uses the foundation of a right to sovereignty to support the idea that individuals have rights over that which they mix their ingenuity, skill and efforts with. This is used to assert that taxation is wrong other than for necessary collective defence and thus entails a vision of sovereignty that not only goes beyond the norm but is hard to square with the common good.

Part of its weakness as a way of thinking is that whilst individual labour deserves credit it is not obvious that it should automatically give rise to almost complete rights of control over that which it is mixed with, in the process necessarily ruling out the idea that rights of control ought to be limited by environmental concerns and also be regarded as a taking from the commons that needs to be compensated for – to wit taxation for the common good would be entirely justifiable.

In sum, the awarding of control and use rights is not a value neutral gesture but rather changes the relationship between all aspects of reality. In other words, those aspects of reality which are awarded the control, those which are subject to the control and those which are not directly party. The nature of that change of relationship can be profound. The global emphasis on the right to property has already, for example, fostered a situation where humans take and keep to a degree and in a fashion that is incompatible with important goals such as enlightened moderation, enjoying and respecting different aspects of reality for their own sake, expressing mutuality and compassion, enjoying equality of opportunity and maintaining sustainability. The critical nature of these goals makes it entirely sensible and reasonable to conceive of them as something that the state and society as a whole are accountable for supporting. Doing so might not fit with the aforementioned libertarian conceptions of property but will increase freedom in its most important sense through enhancing the conditions for human flourishing.

2.3. Justice

The term 'justice' at its broadest refers to the notion that fairness should condition both how individuals, communities and societies behave and the way that structures and systems are created and run by them. One dimension of it that is especially pertinent in the immediate context is the idea that how we treat what I referred to in the previous section as the physicalized and creative dimensions of reality ought to be fair and thus condition, amongst other things, both whether we have a construct of property and, if we do, what rights of control are allowed for under it. It is notable that there are extreme differences of opinion on this matter. Some forms of thought, including Marxism, are predicated on the idea that in so far as rights of control exist they should largely belong in the communal sphere whereas, as I have illustrated in the previous section some, essentially right-wing libertarians, take a diametrically opposed position.

There are also many shades of opinion in the middle including Rawls's view (2001, 42) that, alongside each person having the same indefeasible claim to a fully adequate scheme of equal basic liberties (which scheme is compatible with the same scheme of liberties for all), any social and economic inequalities must satisfy the conditions of being attached to offices and positions open to all under conditions of fair equality of opportunity and be to the greatest benefit of the least-advantaged members of society (the difference principle). Importantly, however, it is very much the norm across the spectrum not to treat rights of control as extending to that which is intrinsic to the human being, including the human body. There are, however, some interesting and often quite contrasting exceptions to the norm. Rakowski (1991) for example, envisages some incursion on the human body as being legitimate to further his agenda of radical egalitarianism but on a scale that is modest relative to some other thinkers. For example, in the transplant context, he envisages compulsory taking of body organs from incompetent minors but otherwise supports the idea of people making voluntary arrangements to enter pools consisting of prospective donors and recipients. Cecile Fabre (2006, 5), by contrast, supports a more extensive degree of incursion with the aid of a much more modest notion of justice – namely, that whilst that individuals have rights to the resources they need in order to live a minimally flourishing life under the principle of sufficiency, once everybody has such a life, all individuals should be allowed to enjoy the fruits of their labour in pursuit of their conception of the good under the principle of autonomy. The positions she advances are twofold: *firstly*, that sufficiency requires the able-bodied to have duties to rescue, provide other personal services and certain body parts imposed on them, with even the latter being capable of being imposed on them without unacceptably compromising their autonomy (2006, 5), *and, secondly*, that sufficiency and autonomy combine to make various need satisfying commercial exchanges with respect to the body acceptable such as in relation to sex, surrogacy and scarce body material provision (2006, 5). Fabre marks out the distinct territory occupied by her position as follows:

> Egalitarian liberals argue that we do not have an exclusive right to the income we get through using our body through labour; in contrast, some libertarians claim that if others have a right against us that we give them part of our income, then, by the same token, they have a right against us that we make

> our body available to them – and that, they think, is a powerful reason to reject coercive taxation for distributive purposes.
>
> Ronald Dworkin's reply to the libertarian argument is a perfect example of egalitarian liberals' strategy at this point: according to Dworkin, we can stave off the libertarian challenge by drawing around the body a 'prophylactic line that comes close to making [it] inviolate.' In this book, however, I argue, against egalitarian liberals and in agreement with libertarians, that being committed to coercive taxation for the purpose of distributive justice does entail that we cannot be committed to a full right to personal integrity.
>
> (2006, 3)

She proceeds to expand on the basis of her position and why she views it as liberal:

> Contra libertarians, I also argue that rejecting the view that individuals have such a right [to personal integrity] does not entail sacrificing one of liberalism's core values – the importance of which underpins many libertarian concerns about taxation – to wit, individual autonomy, and this for two reasons. First, the duty to help by way of providing personal resources is subject to a number of conditions which preserve some space for duty-holders to frame and pursue a meaningful conception of the good. Second, such duty is compatible with the view individuals have the right to sell parts of their body and to lease it for sexual and reproductive purposes, and thereby raise the income they need to implement such conception. By implication, it is also compatible with the view that individuals who are under a duty to make themselves available to others may also have the right to buy body parts and to hire other people's bodies for sexual and reproductive purposes with the latter's consent, and thereby to implement their conception of the good.
>
> (2006, 3)

She goes on to note that

> (t)o the extent that our legal and philosophical tradition forbids the confiscation of personal resources on the grounds that individuals have a fundamental interest in controlling what happens to their person, one might think that it would allow for their commercialization precisely on these grounds.

However, contrary to this,

> our legal and political tradition is such that we have the right to deny others access to our person, even though doing so would harm those who need access; however, we lack the right to use ourselves as we wish in order to raise income, even though we do not necessarily harm others by doing so-even though we might in fact benefit them by doing so.
>
> (2006, 2)

I agree with one implication of Fabre's argument which is that, contrary to the view of certain (essentially right-wing) libertarians (e.g. Nozick, 1974; Narveson, 1983), property rights do not deserve to be shielded from tax or other restrictions designed to foster the common good in some way. However, arguments against having strong property rights cannot be assumed to be translatable into arguments against strong personal rights. The specific idea that it is just to restrict personal rights to satisfy the common good or the specific needs of another or others is difficult to square not just with the holistic vision of reality outlined previously but with liberal democratic thinking in general and, indeed, any approach to governance that includes a significant focus on human rights. Whilst Fabre's exploration of redistribution of interests relating to the body could be considered quite extensive it is much more limited than some. Calabresi (1991), for example, has toyed with the idea of an essentially Marxist system of redistribution to such interests:

> if you are talking about fairness, I really do not understand why the fact that I have inherited good kidneys, or good bone marrow, or good brains, or indeed inherited a good environment, gives me more rights than the person who has inherited bad ones. I am not sure that a person deserves inherited desirable body parts any more than he or she deserves inherited wealth. I might even think the opposite, because with inherited wealth someone else worked so that the person would inherit. The inheritance of body parts, on the other hand, is entirely a matter of luck.
>
> (p. 12)

However, the fact that Fabre's approach manages to leave individuals with some scope to frame and pursue their own conception of the good makes it simply less illiberal rather than liberal. One may also agree with Fabre that sacrificing personal interests on the basis of the needs of others is not incompatible with making commercial gain from the body. However, there are a number of problems with her argument for both.

Firstly, it fails to adopt the kind of critical analysis of demand that would identify the extent, if at all, that real needs could not be met within a system of full protection of personal interests.

Secondly, it sets up a false dichotomy when examining the (general) opposition of philosophical, legal and political traditions to both the sacrifice of person interests and the idea of commercial gain from the body. It is true that elements of these traditions are driven by a liberty based principle of control over the body which makes it hard for them to coherently argue against the ability of the individual to sell the interest in using their body or parts of it. However, equally, it is true that the idea of limits to liberty is invariably recognized within these traditions. Even when those limits are narrowly drawn, they invariably speak partly to the notion that in some cases where it clashes with protection, such as with respect to the incapable, liberty must give way. This harks to the idea advanced in Chapter 1 that respect for human beings is not simply about respecting freedom but also about respecting worth more broadly or what in Western terms is typically described as dignity. As I argued in Chapter 1, the goal of protecting worth broadly provides a coherent basis on which to simultaneously prevent personal rights from being restricted to meet the mere (claimed) needs of others and prevent commercial gain from the human body on the basis that it demeans us both individually and as a species.

Thirdly, whilst it may be argued that a good and fair society needs to foster opportunity for all, the relationship between having resources and flourishing is not so straightforward as Fabre's theory implies and in so far as there is a correlation, it does not follow that there is a right to be able to acquire sufficient resources let alone a right to have them provided. As I have already argued, the treatment of non-human aspects of reality as resource should be constrained by respect for their inherent purpose and value.

The implication of this is that, if their use is ever capable of being justified, it is only going to be justifiable in limited ways and for sufficiently constructive ends. What is constructive in relation to human beings specifically is a lot more subtle than Fabre recognizes. Concerns of a physiological orientation – such as those related to nutrition, hydration, shelter and bodily well-being are not unimportant, but as souls on a journey unlimited by time and space, they must ultimately give way to our more fundamental sense of being and purpose. The limitations of our physiological as well as psychological state are not the product of luck or a vengeful God but something which we can use to evolve through. One may, for example, use the fact of having a disability as an opportunity to evolve certain positive traits – ranging from courage and determination, self-reliance and initiative, to a more acute awareness of the solidarity and lack of solidarity of others. This is not to say that assistance to those in need is always inappropriate, simply that in certain contexts and beyond certain levels it may be counterproductive. Fabre's approach does not allow for this.

Finally, Fabre follows the likes of Huffman (1979) in wrongly conflating the act of an individual choosing not to meet a need with causing harm. To be told one has *caused* harm is effectively to be told that one is *morally responsible* for it. Individuals might be argued to have moral responsibility for leading a good life in general terms but attempts to force them to do so in general terms let alone make specific prescriptions are deeply problematic. One way of criticizing them is to say that they reflect part of what Varden describes as a tendency to think that analysis of 'what is right and wrong from the point of view of ethics or theology will yield the same or at least similar results as an analysis from the point of view of justice' or equally to hold that 'conclusions drawn about what rights we have are dependent on conclusions drawn about what is ethically or religiously required' (Varden, 2016, 35).

I entirely agree with the socially liberal sentiments that lie behind this critique but would put the matter differently by suggesting that the problem lies not so much in the attempt to derive notions of justice or rights out of ethics or religion but out of forms of them which are degrading and/or repressive with Fabre's argument being the former because of its commodificatory implications and both

Huffman's and Fabre's arguments being the latter for want of due respect for liberty. Trying to impose a specific notion of the good life is not pragmatic either. All other things being equal an individual will know what is a good approach to life for her better than others will, she will learn partly through thinking about how she can improve on what she has previously done and allowing others to decide for her is to take the risk that they may abuse their power or in any event decide no better than she would and in some cases worse. Beyond providing certain protections against individuals significantly damaging others or themselves, our efforts should not go beyond encouraging each other to think and live holistically and thus in a way that is likely to be harmonious with growth. Defining individuals as morally, let alone legally, culpable for turning down specific demands for their use is a grossly illiberal and ultimately extremely destructive invasion of their right to explore, define and live their own vision of the good life.

Another line of argument that is essentially justice based is that there is a moral duty to contribute that can legitimately be translated into law. This is sometimes framed as general (English et al., 2006; Brownsword, 2008) and sometimes framed as an argument relevant to a specific context involving medical use of the human being, such as medical research (Harris, 1997; Gunn et al., 2000) or transplantation (Merle, 2000). It is often built on two points: firstly, the idea that human beings have a duty of virtue and, secondly, the idea that this should be treated as enforceable in various ways. I expressed qualified support for the first view in the previous chapter on the basis that human beings have an inherent purpose and drive toward growth. However, I also emphasized that the path of growth for each individual is only fostered by high levels of respect for autonomy, and clearly this is inconsistent with forcing things on them that are against their bodily interests in order to meet the demands or needs of others. Merle (2000) tries to make the argument, in the organ donation context, that Kant would have been in support of it but, even leaving aside questions of whether Kant would have supported organ donation at all (Gerrand, 1999), this seems to run counter to Kant's assertion that the duty of virtue was not a categorical imperative.

The argument for a compelled duty to contribute is sometimes bolstered by an appeal to the idea health law is too individualistic and insufficiently communitarian both generally (e.g. English et al., 2006; Brownsword, 2008) and relatedly by the argument that there is a social agreement or contract between the individual and the state (e.g. English et al., 2006). Part of the appeal of these arguments is that many societies are indeed too individualistic. More reciprocity, solidarity and care are needed to make them work better. However, diluting protection of body related rights of individuals to meet claimed needs for medical use is not an acceptable and effective way of increasing emphasis on them. They are values that can only be freely generated. They can be fostered when human beings are engaged with in an open and unbiased way but not when they are being treated as mere means such as by being mandated or manipulated to contribute or in any other way having their normal legal protections diluted. Furthermore, when dilution on this basis is treated as legitimate, there is evidence both from within discourse, practice and governance that it becomes an excuse for selective disadvantageous treatment of certain classes of human being, i.e. for discrimination. Those who are particularly 'useful' or weak or vulnerable for organic or socially constructed reasons find themselves targeted.

Potentially more legitimate territory comes in the form of the argument that reciprocity at least can be used to form a link between the provision or willingness to provide healthcare 'goods' and the ability to obtain them. Such a link would be anti-pluralistic if imposed in a generalized way but might be considered more acceptable if it was hypothecated in such a way as to only interfere with people's ability to be hypocritical. Gunn et al. (2000, 63) give the example of those who want to benefit from research being obliged to offer themselves for participation to avoid being a parasite or freeloader. Harris expresses a similar view (1997). However, both of these authors use their sentiments purely as a basis to argue for a weakening of protections in research rather than to support hypothecation. Furthermore, research is often unsuited to hypothecation since it is hard to link participation in a study of a particular area or condition to the obtaining of benefits that relate to that and would be oppressively broad to deny people access to the benefits of research as a whole consequent on their non-participation.

Transplantation, by contrast, is potentially much more suitable. The ability to obtain an organ, for example, could be linked to the registration of willingness to become an organ donor. Rupert Jarvis's modest version of this is that those who have registered to donate an organ or organs after their death should in the event of subsequently needing an organ during life, be prioritized on any waiting list which exists for that type of organ (Jarvis, 1995). In the interests of fairness, such a proposal might need to be limited to those who have capacity and have resided in the country in question for a reasonable period of time. However, even with these and other conceivable provisos, it might have the effect of increasing the pot of available organs and thereby decrease waiting lists overall. Frenkel (1975) had earlier suggested a much broader approach to hypothecation involving patients needing an organ being prioritized for it where one of their first degree relatives (spouse, child, parent or sibling) had already become a cadaveric organ donor or if they or such a relative had signed up for cadaveric donation of an organ or organs (but not mere tissues) at least 3 years earlier or have been a living kidney, lung lobe or liver segment donor. Israel's Organ Transplantation Law 2008 includes this very system. Quigley et al. (2012) have suggested that the inclusion of first degree relatives makes it too easy to gain priority. Wood and Frenkel have countered by suggesting that the first degree relative element has been an important factor in the subsequent growth in numbers of people signing donor cards (2014, 172).

Those who believe transplantation to be an appropriate focus for hypothecation might still wish to reject it in the above forms on the basis that it is incompatible more generally with the bond of trust that is supposed to be central between healthcare providers and professionals on the one side and the public they serve on the other.

2.4. Utilitarianism

There is a natural appeal to considering whether ends are useful to pursue, before pursuing them. However, utilitarianism goes beyond being a set of part overlapping and part competing visions about what is optimally useful and how to achieve it. It is a framework that *prioritizes* what is considered useful above everything else. It is often considered to have been developed by the English thinker Jeremy Bentham (1748–1832), but his work was acerbically described by Karl Marx as simply reproducing 'in his dull way what Helvétius and other Frenchmen had said with esprit in the 18th century' (1863, 668 ft 2). Furthermore, the roots of utilitarianism undoubtedly go much further back still – as John Stuart Mill (1806–1873) noted it was element of ancient Greek thought (1998, 131, 136). Nonetheless, Bentham's work certainly brought about a renewed emphasis on it.

Both he and subsequently Mill embraced a 'Greatest Happiness Principle' conception of it which emphasized actions as right solely in proportion to the extent that they produce happiness and wrong solely in proportion to the extent that they produced its opposite which they rather strangely viewed as pain. They saw the ultimate goal as being to organize society in such a way as to produce the greatest happiness for the greatest number of people. In accounting for the juxtaposition of utility and pleasure, Mill emphasized that utilitarians have always stressed that utility is

> not something to be contradistinguished from pleasure, but pleasure itself, together with exemption from pain; and instead of opposing the useful to the agreeable or the ornamental, have always declared that the useful means these, among other things.
>
> (1998, 36)

The emphasis on pleasure is unfortunate because it is suggestive of a self-destructive and facile hedonism and has the capacity to be used in this way instead of in a manner that is actually consistent with utility. It is helpful in this context that modern utilitarians tend to eschew focusing on pleasure for a more rounded and robust focus of well-being. Furthermore, some utilitarians have suggested that there ought to be rules which one generally follows even in a case by case analysis of well-being. With respect to individuals, such rule based utilitarianism can be focused around the fact that there are certain skills, attributes and states which are conducive for well-being and/or development. With respect to societies, it can reflect the fact that there are certain social and political arrangements, and even values and norms, which it might generally be useful to protect, There have also been efforts, most notably from Peter Singer (1975), to frame utilitarianism in a manner that goes beyond human beings to account for animals.

However, such efforts should not detract from the fundamental flaw that utilitarianism has when advocated as the directing principle for society as a whole. Its very nature, even in rule based forms, is to 'subjugate everything . . . to the end of maximizing overall utility' (Garwood-Gowers, 2013, 396). By *everything* in this context, I mean both every aspect of reality and also every potential other end than itself. I reject even the idea that utilitarianism promotes the group and the community since I consider, as noted in Chapter 1, the supposed tussle between individual and wider interests to be grounded on a false dichotomy. More particularly, I am arguing that on its group and societal levels, utilitarianism is inherently counterproductive: One can never actually maximize well-being within a society governed through the principle that its maximization requires an open hand in dealing with the individual because it leaves all of the individuals that make up society in a constantly vulnerable and defensive position, unable to flourish because they are always wondering what might be done to them.

In reflecting on this in more detail, it is instructive to return to Bentham's attempt to defend his work partly through criticism of the emphasis on natural law that had emerged over the prior centuries. Bentham singled out the work of Grotius and Pufendorf for particular ridicule, suggesting that they were trying to legislate for the world by arm chair and conquer it by pen with their 'idlest trash' being considered an oracle and this oracle being considered a law (1843, Vol. 3, 221). Bentham viewed the idea that natural rights could be derived from natural law to be 'simple nonsense' and considered natural and imprescriptable rights to be 'rhetorical nonsense, – nonsense upon stilts' seemingly specifically because no government could on any occasion abrogate 'the smallest particle' of them (1843, Vol. 2, 502). Whilst there are legitimate reasons to debate the grounding and precise scope of rights, Bentham's evident tone of scorn for the idea of protecting them in general can be seen as somewhat ironic, since precisely what nonsense is it to ask us to have trust in a state that is left with a single duty to maximize utility? Especially when one considers that this duty is so amorphous in character as to be ineffectual in placing any real limits on how power is held and exercised. Indeed, advocacy of utilitarianism advances the concentration and abuse of power, even to the point of tyrannous dictatorship. This may be one factor in what Masferrer (2016, 228) has described as a decline in utilitarian scholarship since the end of the nineteenth century. Certainly, World War 1 and World War II made a lot of scholars think harder about the actual real world consequences of unconstrained exercise of state power and, as we have seen, ultimately activated an impulse to contain such power, not least through the protection of human rights. The propagation of views of the good of the whole has, however, not merely been leveraged for tyranny by dictators but also within more regular democratic regimes. As Masferrer notes:

> Regarding natural rights as 'nonsense', or fundamental rights as the mere creatures of the state that enforces them with no more legitimacy than the binding force of the law (understood as a mere command of the state), seems to be the best 'foundation' for the violation of rights whenever the state finds them inconvenient for whatever reasons.
>
> (2016, 229)

As an instrument of concentration of power, utilitarianism tends to defend established order and related orthodoxy. In medicine this means eschewing a holistic approach with an emphasis on prevention and cure for an industrialized and mechanistic approach that, amongst other things, relies excessively on medical use to achieve its goals.

Attempts have been made to soften its effect by trying to combine it with a deontological approach. For example, Morrison (2005) has tried this in the context of clinical research. However, the very nature of even enlightened forms of utilitarianism is that they will not allow the objective of maximizing utility to be traded off against anything. Equally, deontological thought by its nature treats certain values or principles as absolute and thus incapable of being traded off. The two approaches are in this sense fundamentally opposed. The point can perhaps best be illustrated through considering the work of John Stuart Mill. Largely in the context of law and society in mid-nineteenth century Britain, Mill spoke out about such issues as the suppression of women, the need for representative government and the social and legal importance of liberty more generally. His overall emphasis on liberty is often summed up by reference to his statement that:

> The sole end for which mankind are warranted, individually or collectively, in interfering with the liberty of action of any of their number is self-protection. That the only purpose for which power can be rightfully exercised over any member of a civilized community, against his will, is to prevent harm to others. His own good, either physical or moral, is not a sufficient warrant. He cannot rightfully be compelled to do or forbear because it will be better for him to do so, because it will make him happier, because, in the opinions of others, to do so would be wise, or even right. These are good reasons for remonstrating with him . . . but not for compelling him, or visiting him with any evil in case he do otherwise. To justify that, the conduct from which it is desired to deter him, must be calculated to produce evil to someone else. The only part of the conduct of any one, for which he is amenable to society, is that which concerns others. In the part which merely concerns himself, his independence is, of course absolute. Over himself, over his own body and mind, the individual is sovereign.
>
> (1998, 14)

Mill sometimes appears to be a closet libertarian trying to wriggle free of the utilitarian straightjacket passed down by his father and Bentham. However, ultimately his support for liberty is relative to circumstances rather than fundamental. He makes this evident in the context of freedom from forced action by arguing that people should be legally obligated

> to perform certain acts of individual beneficence, such as saving a fellow creature's life, or interposing to protect the defenseless against ill-usage . . . there are often good reasons for not holding him to the responsibility; but these reasons must arise from the special expediencies of the case: either because it is a kind of case in which he is on the whole likely to act better, when left to his own discretion, than when controlled in any way in which society have

it in their power to control him; or because the attempt to exercise control would produce other evils, greater than those it would prevent.

(1998, 14)

Such a willingness to compel rescue in the name of utility is also evident within some of the ideation that is more specific to medical use. A classic and extreme illustration is the idea put forward by the philosopher John Harris that we should have a survival lottery involving everyone being entered into a regular random draw with those who 'number came up' being forced to give up certain body materials to save the lives of others (Harris, 1975). Harris suggested that the conscription could be extended to forcing those randomly selected to give up organs that were necessary to maintain their own life so as to maximize the utility benefits to others.

To sum up, utilitarianism is politically naïve and dangerous in the way that it facilitates the misuse of power, not least in the context of medical use. Its emphasis on trying to maximize happiness is a poor and superficial substitute for developing a proper vision of how to protect worth that is grounded on an understanding of nature and purpose. In the human context specifically, such a vision should necessarily be seen as incorporating various principles and values, including some that need to be protected absolutely. Far from being nonsense on stilts, the need to have a schema of rights and protections naturally flows from this.

A variation on the theme of utilitarianism is to regard utility as one of several ends to be pursued. It is easy to support this as a basic idea but equally important to recognize that it too can be misused as a tool to bring about interventions that are not consistent with respect for human beings. Indeed, this is exactly what frequently happens in the medical use context, particularly with respect to vulnerable groups such as the incapable adult.

2.5. Conclusion

This chapter confirms the view posited in the introduction and Chapter 1 that ethical or moral arguments can be divided into those which are broadly compatible with respect and those which are not, including theories underpinned by notions of one or more of liberty, justice and utility. I would agree with the Western liberal tradition that liberty is a vital underpinning of democracy and integral to respect not only in as much as it leaves individuals free to pursue their goals but also because this in turn is integral to their development. However, equally I have referred to the potential for an excessive form of individualism which tends towards disintegration. This may be said to be borne out of what Midgley (1994, 123) describes as the tendency within Western cultures for the degree to which human individuals are independent, separate and internally harmonious organisms to be grossly exaggerated. Nonetheless, a respect based approach rather than one that dilutes respect to achieve other ends is necessary to give effect to what Midgley describes as the modest facts that underlie individualism. It acts as a buffer against intrusions on personal rights that are derived from excessive and ultimately counterproductive forms of communitarianism. At the same time, where grounded on a more rounded concept of worth than one simply based on liberty, it prevents the capacity of individuals for agency to be used as a basis to defeat the legitimate concern of the community to protect its members from exploitation.

References

Bentham, J. (1843a) *The Works of Jeremy Bentham* (Vol. 2). Available at http://oll.libertyfund.org/titles/1921#Bentham.

Bentham, J. (1843b) *The Works of Jeremy Bentham* (Vol. 3). Available at http://oll.libertyfund.org/titles/1922#Bentham.

Brownsword, R. (2008) *Rights, Regulation, and the Technological Revolution*. Oxford: Oxford University Press.

Calabresi, G. (1991) Do we own our own bodies. *Health Matrix*. 1(5), 5–18.

Cambridge Online Dictionary. (2019) Available at https://dictionary.cambridge.org/dictionary/english/libertarian.

English, V., et al. (2006) Autonomy and its limits: What place for the public good? In McLean, S. (ed.) *First Do No Harm: Law, Ethics and Healthcare*. Aldershot: Ashgate, pp. 117–130.

Fabre, C. (2006) *Whose Body Is it Anyway? Justice and the Integrity of the Person*. Oxford: Oxford University Press.

Frenkel, D. (1975) *The Legal Aspects of Human Organ Transplants*. LLD Thesis. Hebrew University of Jerusalem.

Garwood-Gowers, A. (2013) Time to address the problem of post-mortem procurement of organs for transplantation occurring without proper pre-mortem consent. *European Journal of Health Law*. 4, 383–408.

Gerrand, N. (1999) The misuse of Kant in the debate about a market for human body parts. *Journal of Applied Philosophy*. 16, 1.

Gunn, M., et al. (2000) Medical research on incompetent adults. *Journal of Mental Health Law*. 60–72.

Harris, J. (1975) The survival lottery. *Philosophy*, 81–87.

Harris, J. (1997) The ethics of clinical research with cognitively impaired subjects. *Italian Journal of Neurological Sciences*. Supplement, 5, 9–13.

Huffman, F. (1979) Coerced donation of body tissues: Can we live with McFall v Shimp? *Ohio State Law Journal*. 40, 409–440.

Jarvis, R. (1995) Join the club: A modest proposal to increase availability of donor organs. *Journal of Medical Ethics*. 21, 199–204.

Laslett, P. (1960) *John Locke's Two Treatises of Government*. Cambridge: Cambridge University Press.

Machan, T. (2006) *Libertarianism Defended*. Aldershot: Ashgate.

Mackenzie, C., et al. (2014) Introduction: What is vulnerability and why does in matter for moral theory. In Mackenzie, C., et al. (eds.) *Vulnerability*. Oxford: Oxford University Press, pp. 1–29.

Marx, K. (1863; 2007) *A Critique of Political Economy: The Process of Capitalist Production as a Whole* (Vol. 111(11)). New York, NY: Cosimo.

Masferrer, A. (2016) Taking human dignity more humanely. In Masserer, A. & Garcia-Sánchez, E. (eds.) *Human Dignity of the Vulnerable in the Age of Rights*. Heidelberg: Springer, pp. 221–256.

Merle, J-C. (2000) A Kantian argument for a duty to donate one's own organs: A reply to Nicole Gerrand. *Journal of Applied Philosophy*. 17(1), 93.

Midgley, M. (1994) *The Ethical Primate*. Abingdon: Routledge and Kegan Paul.

Mill, J. S. (1998) *On Liberty and Other Essays*. Oxford: Oxford University Press.

Morrison, D. (2005) A holistic approach to clinical and research decision-making: Lessons from the UK organ-retention scandals. *Medical Law Review*. 13, 45–79.

Narveson, J. (1983) On Dworkinian equality. *Social Philosophy and Policy*. 1, 1–23.

Nozick, R. (1974) *Anarchy, State and Utopia*. New York, NY: Basic Books.

Quigley, M., et al. (2012) Organ donation and priority lists in Israel. *Transplantation*. 93, 970–973.

Rakowski, E. (1991) *Equal Justice*. Oxford: Oxford University Press.

Rand, A. (1957) *Atlas Shrugged*. New York, NY: Penguin.

Rawls, J. (2001) *Justice as Fairness: A Restatement* (ed. Kelly, E.). Cambridge, MA: Belknap Press.

Singer, P. (1975) *Animal Liberation: A New Ethics for our Treatment of Animals*. New York, NY: New York Review; Random House.

Varden, H. (2016) Self-governance and reform in Kant's liberal republicanism – Ideal and non-ideal theory in Kant's Doctrine of Right. *Curitiba, São Carlos*. 13(2), 39–70.

Ward, C. (1988) *Anarchy in Action*. London: Freedom Press.

Wood, D., and Frenkel, D. (2014) Reflections on the Israeli legislation on transplantation. In Weimar, W., et al. (eds.) *Organ Transplantation: Ethical, Legal and Psychosocial Aspects: Global Issues, Local Solutions*. Lengerich: Pabst, pp. 167–173.

3 Human rights based constraint

3.1. Introduction

Protecting human worth and protecting human rights are not entirely analogous activities, but they do certainly heavily overlap. It is thus entirely apposite to consider the extent to which constraint of medical use has and could be achieved through human rights law and its enforcement. Since, medical use normally comes via the body, the central concern in this focus in the chapter is on the protection of rights in or related to the body, specifically from intentional intrusion. In many jurisdictions, this has a long tradition of being achieved through criminal and civil law as well as a typically somewhat more recent tradition of being protected via public law. In English law, for example, criminal law protections of the person from intentional harm (extending all the way up to homicide) have existed for thousands of years, whilst civil ones (trespass to the person) can be traced to at least early medieval times (see more specific discussion within Chapter 3) at which point public law ones also began to emerge, such as via the Magna Carta 1215. Over time the range of body related interests that exist has started to become better elucidated both in the general law and within public law more specifically at the domestic level. However, this chapter will focus largely on global and regional developments rather than domestic ones specifically, reflecting the fact that its aim is to address the macro question of how far human worth is generally being protected, rather than the more micro question of how it is being protected within individual jurisdictions. I will specifically argue that general and health specific human rights provisions at the global and regional levels are structured in such a way that the protection of human worth in the context of medical use is for the most part an imperative.

A key initial impetus for the development of global and regional human rights instruments was a growing feeling, following the end of World War II, that something must be done to prevent a return to the scourge of war and also to also more specifically provide a counterweight to the threat near unlimited state power posed to the worth of the individual, as vividly illustrated by the atrocities that had occurred during the course of the war, not least the Holocaust. One of the key developments in this regard was the coming together of a group of what were then 11 nations at the end of the War to form the United Nations Charter 1945. Not long after that important general human rights instruments were developed by both the United Nations and a number of regional bodies. Reflecting the fact that many of the atrocities committed during the war were medical in character (see further Chapter 4), there were also more specific efforts to better contain the exercise of medical power. In more recent decades, there have also been further waves of development of general and health specific human rights instruments – the latter particularly sparked by movements concerned with protecting personal freedom and minimizing the abuse of medical power, not least in the context of biotechnology.

The principal aims of this chapter are to evaluate the extent to which the total web of relevant provision at the global and regional levels sets a moral, political and even legal tone of supporting respect for human beings in the context of medical use and to critically examine what reforms to it might enable it to do so more effectively. Some domestic systems are also briefly examined by way of considering the ways in which respect does and/or does not permeate into the approach that they take to medical use.

3.2. Global instruments

UN general human rights instruments

The Universal Declaration of Human Rights (UDHR), 1948, and *The* International Covenant on Civil and Political Rights (ICCPR), 1966, are widely recognized as two of the most important human rights ever created. The former has been described as 'the cornerstone of the human rights system' being 'the first universal statement of the basic principles of inalienable human rights, and a common standard of achievement for all peoples and all nations' (Andorno, 2016, 264). The latter builds on it, not just in terms of the scope of its rights and protections, but also through the fact that it has treaty status and is enforced through the UN Human Rights Committee (UNHRC). The UNHRC not only monitors compliance by individual states but provides a mechanism for individuals to complain about violations of its provisions.

Both instruments emphasize dignity (UDHR, Art. 1, and ICCPR, preamble) and stress that the rights within them apply to all without discrimination (UDHR, Art. 2, and ICCPR, Art. 2(1)). They also have a considerable range of protections that can protect interests engaged in the process of medical use of the body. The right to private and family life (ICCPR, Art. 12, and UDHR, Art. 17) is the most generically relevant on these since it applies to all situations in which physical integrity is at stake (see, for example, *VDA v Argentina* [1608/2007] UN Human Rights Committee 29 March 2011, 30 B.H.R.C. 531). Meanwhile, protection against torture, inhuman or degrading treatment (UDHR, Art. 5, and ICCPR, Art. 7) is relevant in selective circumstances as are protection against medical or scientific experimentation without free consent (ICCPR, Art. 7) and of the right to life (UDHR, Art. 3, and ICCPR, Art. 6).

Whilst not a general human rights instrument, the 2006 Convention on the Rights of Persons with Disabilities (CRPD) is also worth mentioning because of its impact on furthering respect for bodily interests in the disability context. Perhaps the CRPD's most significant overall feature is that it embodies in the disability context a trend toward broadening of the scope of the right to equality and non-discrimination 'away from a strictly formal towards a more substantive notion of the right to equality, both in the formulation of the norms themselves and in their application by (semi) judicial bodies' (Loenen, 2001, 210). It has a number of provisions which apply it emphasis on equality in the disability context to how the body is treated. For example:

- Art. 15 of the CRPD applies the emphasis of the Covenant and Declaration on freedom from torture, inhuman and degrading treatment and the emphasis of the latter of not conducting medical or scientific experimentation on people without free consent to the disability context.
- Art. 17 more broadly protects against discriminatory intrusion on the integrity of this class in the medical context and beyond by recognizing their equal right to respect for their physical and mental integrity.
- Art. 16 goes beyond this by protecting their freedom from all forms of exploitation, violence and abuse.
- Art. 25 prevents discriminatory denial of healthcare or health services or food and fluids on the basis of disability.
- Art. 10 protects the right to life in the disability context.

These provisions are very considered and robust, though the right to life provision has rightly been criticized by, amongst others, Kayess and French (2008, 29) for remaining silent on 'genetic science aimed at the elimination of impairment-related human diversity, and on pre-birth negative selection of foeti with identified or imputed impairment.'

The Nuremberg Code and World Medical Association provision

The ten principles for conducting human experimentation issued as part of the judgment in the so-called Doctors Trial (*United States of America v Karl Brandt et al. 1946–7*, part of the Nuremberg Nazi War Crimes Trials), which became known as the Nuremberg Code. The Code is notable in the immediate context for the way in which it reflects a respect for worth or dignitarian balancing of the autonomy of prospective research subject and their protection from misuse. Specifically, it requires participation to be voluntary and informed in the fullest sense (Principle 1) and capable of being withdrawn at any time (Principle 9), but also provides that research needs to be well designed; have aims that cannot be achieved by other means and worthwhile potential results (Principles 2 and 3); involve risk that is not excessive or disproportionate to benefit (Principles 5 and 6); and be conducted in a manner protective of well-being (Principles 4, 7, 8 and 10).

The imperative of better aligning medical practice as a whole to respect was to some extent taken on by the newly developed World Medical Association (WMA) in this period.

The Association's *Declaration of Geneva Physicians Oath 1948* (amended most recently at the 173rd Council Session in May 2006 at Divonne-les-Bains, France) requires medical professionals to swear, amongst other things, that they will practice with 'conscience and dignity, treat the health of their patients as their 'first consideration' and avoid using medical knowledge 'to violate human rights and civil liberties' (originally worded 'contrary to the laws of humanity'). Its *International Code of Medical Ethics 1949* (as amended most recently in October 2006 at the General Assembly in Pilanesberg, South Africa) requires physicians to 'act in the patient's best interest when providing medical care.'

Perhaps the WMA's most important contribution has been the development of a detailed instrument for the regulation of medical and scientific research on human beings – namely, the *Declaration of Helsinki.* Since its adoption in 1964, the Declaration has been amended many times, most recently in October 2013 at the 64th WMA General Assembly in Fortaleza, Brazil. The Declaration differs from the Code in a number of respects that are important in the immediate context.

Firstly, in terms of scope, it covers therapeutic as well as non-therapeutic research.

Secondly, it is more detailed and contains more procedural controls (including the requirement for independent ethical review committees introduced in 1975).

Thirdly, it contains a specific and unequivocal emphasis on the primacy of the individual as a consequence of Principle 8 which states that, '(w)hile the primary purpose of medical research is to generate new knowledge, this goal can never take precedence over the rights and interests of individual research subjects.' This is complemented by Principle 7 which emphasizes respect for subjects and protection of their health and rights and Principle 9 which emphasizes protection of their life, health, dignity, integrity, right to self-determination, privacy and confidentiality of personal information.

Finally, its specific provisions on the incapable envisage them being used as participants in research. Indeed, they do not require such use to benefit let alone be in the best interests of the incapable (see particularly Principle 26), although both these should be deemed to be mandated by its more general emphasis on primacy.

The Council of Europe's Convention on Human Rights and Biomedicine (CHRB)

Whilst the CHRB was created by the Council of Europe it is best classified as a global instrument in an as much as non-European states can and in some cases have signed up to it. Its respect based focus is evident from its aim being to recognize and protect human rights and dignity in the context of medical and biological intervention on human beings. It is possible to identify four normative pillars within it that flow from this aim: firstly, 'protection of the dignity and the identity of the human being . . . in line with other international documents'; secondly, 'the concept of integrity . . . [which] embraces both the right to privacy and the right to inviolability of one's body'; thirdly, 'the right of equal access to healthcare'; and, fourthly, 'non-discrimination' (Dute, 2005, 8). The pillars most relevant to medical use – namely the first, second and fourth – are emphasized in a foundational manner in chapter 1 and then applied in the six following chapters which cover, in turn: consent (chapter 2), private life and the right to information (chapter 3), the human genome (chapter 4), research (chapter 5), transplantation (chapter 6) and non-commodification with respect to the body (chapter 7). Within chapter 1 the critical provision is Art. 1 which states that parties to the CHRB

> shall protect the dignity and identity of all human beings and guarantee everyone, without discrimination, respect for their integrity and other rights and fundamental freedoms with regard to the application of biology and medicine.

In working through this provision one can note that some rights and freedoms necessarily have to be treated as qualified. The same is true of integrity. However, dignity constitutes the CHRB's essential value to be upheld (Explanatory Report, para. 9), and from this one can conclude that it is absolute. According to the schema developed in chapter 1, protecting the identity of human beings and equality of treatment as between them can be seen as functions of protecting dignity and accordingly absolutely protected by Art. 1 (even though the latter is only specifically highlighted in the context of respect for integrity and other rights and fundamental freedoms).

When Art. 1 is read in this fashion, it is open to debate whether Art. 2 adds anything by stating that '(t)he interests and welfare of the human being shall prevail over the sole interest of society or science.' Dute (2005, 8) asserts that the function of Art. 2 is to ensure that human beings are treated as ends rather than mere means. However, whilst this is consistent with its title ('Primacy of the human being'), it does not fit well with the fact that the text only explicitly talks about protecting human beings in the context of scientific and societal interests and not in the case of one individual's interests conflicting with those of another (Gadd, 2001, 26). However, even if it does not prevent instrumental use of one individual to fulfil the (claimed) needs of another, Article 1 will as a result of its emphasis on dignity. The need to avoid such treatment is also validated by the scope of the limits that Art. 26 places on the extent to which the rights, freedoms and protections laid out in the CHRB as a whole can be restricted. Art. 26 states that:

> (1) No restrictions shall be placed on the exercise of the rights and protective provisions contained in this Convention other than such as are prescribed by law and are necessary in a democratic society in the interest of public safety,

for the prevention of crime, for the protection of public health or for the protection of the rights and freedoms of others; and

(2) The restrictions contemplated in the preceding paragraph may not be placed on Articles 11, 13, 14, 16, 17, 19, 20 and 21.

Art. 26(2) mandates that parties should treat the rights and protections contained in it as absolute. Meanwhile, Art. 26(1) closely circumscribes the scope that they have to restrict others by requiring such restrictions to be prescribed by law and necessary in a democratic society for the protection of one of the four interests laid down in it. Restrictions which use human beings as mere means would be at odds with – rather than necessary in – a democratic society. This is perhaps why the examples of potentially acceptable restrictions listed in the Explanatory Report all relate to situations in which the conflicting rights rather than mere needs of others are at stake. Specifically:

- Para. 150 lists '(c)ompulsory isolation of a patient with a serious infectious disease, where necessary, is a typical example of an exception for reason of the protection of public health';
- Para. 151 states that non-consensual confinement and treatment of a person with mental disorder may be justified not just in cases Art. 7 contemplates but also to prevent possible serious harm to others under the rights and freedoms of others element of Art. 26(1);
- Para. 152 notes that this provision may also justify an order by a judicial authority for a test to be carried out to establish parentage;
- Para. 153 suggests that genetic assessments (DNA tests) may be justified for the identification of persons in connection with criminal investigation; *and*
- Para. 154 suggests that Court ordered treatment may be compliant with Art. 26 in some circumstances.

At the same time, not all of the CHRB's more issue specific provisions sit easily with the four absolutes protected in Articles 1 and 2. Shortfalls in the way medical use tailored provisions operate and how these could end up being mitigated are summarized as follows.

Firstly, shortfalls in its approach to commodification. The CHRB prohibits the human body and its parts being, as such, the subject of financial gain through Art. 21. However, para. 134 of the Explanatory Report makes it clear that this does not extend to issues of patentability and patenting of biotechnological inventions. The commodificatory effects of such patenting stem from the fact that it encourages people to violate body protective norms in order to profit and the fact that it allows for a degree of profit to be made which far exceeds reasonable remuneration for effort, ingenuity, skill, expense and even commercial risk. The famous case of *Moore v Regents of University of California* 793 Pd 479 (1990) provides a window through which one can clearly view both of these problems. Here a cell line developed from materials taken from the claimant John Moore was valued at an estimated $3.01 billion by 1990. The creation of the cell line rested on a considerable amount of effort, ingenuity, skill and expense, but its high value rested largely on the happenstance of the particular qualities of Moore's body. Moore had consented to the removal of these materials for diagnostic and therapeutic purposes but had no inkling that they would be commercially exploited. The limitations of Art. 21 could be addressed through its reform but in the meantime could still be addressed through relying on Art. 1, since its absolute protection of human dignity implies a need to entirely oppose commodification. Art. 1 could thus, for example, be used to mandate that organizations should only be able to benefit from patenting of biotechnological inventions where they operate on a not-for-profit basis.

This leads us on to the *second* point that the CHRB's approach to storage and use of body material for a purpose other than that for which it was removed is also deficient. Art. 22 states that such storage and use must conform with appropriate information and consent procedures. However, paragraph 137 of the Explanatory Report takes a limited view of what this entails, treating limited provision of information as potentially adequate, equating non-opposition by patients or their representatives with consent and suggesting that the necessity for express and specific consent is selective without precisely identifying situations in which it would not be required and why. Taken in isolation, it treats the individual as a mere means to the end of meeting medical demands in these respects. So does Art. 14 of the 2008 Additional Protocol Concerning Genetic Testing for Health Purposes by allowing some scope for genetic tests to be performed on the previously removed biological materials of a person who cannot be contacted for the benefit of his or her family 'in accordance with the principle of proportionality, where the expected benefit cannot be otherwise obtained and where the test cannot be deferred.' It should be read down in the light of the fact that the AP also emphasizes individual primacy (Art. 1 and 3) and the main Convention emphasizes both primacy and dignity (Art. 2 and 1).

Thirdly, the CHRB is also patchy in applying the principle of equality in certain specific contexts, including some which involve or could involve medical use. Art. 11 provides absolute protection against discrimination on grounds of genetic heritage and Art. 14 provides absolute protection against sex selection in medically assisted reproduction, except where serious hereditary sex-related disease is to be avoided. However, whilst Art. 18 prohibits the creation of in-vitro embryos for research purposes, it allows research on existing in-vitro embryos subject to the unexplained and perhaps virtually meaningless requirement that they are adequately protected. Its research provisions also do nothing to prevent research on the incapable that is inconsistent with their best interests. Indeed, Art. 17 actually encourages such research. By contrast, the Additional Protocol on Biomedical Research would appear to oppose this (despite Art. 15 of it replicating Art. 17 of the CHRB). Its preamble expresses the convictions 'that biomedical research that is contrary to human dignity and human rights should never be carried out,' that 'the paramount concern' is 'the protection of the human being participating in research' and 'that particular protection shall be given to human beings who may be vulnerable in the context of research.' Furthermore, Art. 3 of it states that, '(t) he interests and welfare of the human being participating in research shall prevail over the sole interest of society or science.' The extent to which the Convention extends to the not yet born and the dead is uncertain.

Fourthly, in the context of genetic intervention, the CHRB partially departs from its Art. 1 commitment to protect identity. Its Art. 13 prohibition on interventions 'seeking to modify the human genome' does not extend to those 'undertaken for preventive, diagnostic or therapeutic purposes' where their 'aim is not to introduce any modification in the genome of any descendants.' The Council of Europe treads a fine and debatable line here and in its Additional Protocol on the Prohibition of Cloning of Human Beings 1998, treating cloning as contrary to dignity in cases where the clone is implanted but not where it is merely created.

Taken as a whole, it seems the limitations of some of the issue specific provisions of the CHRB in the context of medical use can be addressed by referring back to the norms it protects as absolutes via Art. 1. Furthermore, even if that were not so in certain instances, the CHRB would be guilty of colluding with shortfalls rather than mandating them since as Art. 27 makes clear, it operates as a floor rather than a ceiling on rights and protections. It is also notable that by virtue of Art. 36 reservations made by States when signing the CHRB or depositing the instrument of ratification, acceptance, approval or accession can only be with respect to particular provisions not those of general character. This would seem to exclude reservations at odds with the norms it protects as absolutes via Art. 1.

UNESCO bioethics related provision

UNESCO has been undertaking work in the bioethics field for over two decades and has produced two significant instruments in this period: the Universal Declaration on the Human Genome and Human Rights (UDHGHR) in 1997 and the Universal Declaration on Bioethics and Human Rights (UDBHR) in 2005. Whilst the UDHGHR is focused on human genome science, the UDBHR applies more broadly to medicine, life sciences and associated technologies as applied to human beings, making it more similar to the CHRB in terms of the breadth of its scope. However, all three instruments are broadly similar in their basic normative orientation, although they differ significantly in their precise manner of expressing that orientation.

The UDBHR is designed, amongst other things, to 'respect for human dignity and protect human rights, by ensuring respect for the life of human beings, and fundamental freedoms' (Art. 2). This design is reflected in foundational terms in the first of its substantive articles (Art. 3) which stresses that '(h)uman dignity, human rights and fundamental freedoms are to be fully respected' and then goes on to ape the CHRB's Art. 2 emphasis on primacy of the human being. Several later articles stress equality as between human beings and related notions of non-discrimination and protection of the vulnerable (see, especially, Art. 8, 10 and 11).

The UDHGHR if anything has a more predominant emphasis on dignity than the UDBHR and perhaps even the CHRB, no doubt partly because one is very directly and potently confronted with threats to dignity in the context of human genome science. Art. 1 of the UDBHR emphasizes that the human genome underlies the fundamental unity of all members of the human family, recognizes their inherent dignity and diversity and is symbolically the heritage of humanity. Art. 2 applies this by stressing that everyone has a right to respect for their dignity and for their rights regardless of their genetic characteristics (2a) and emphasizing that such dignity makes it imperative not to reduce individuals to their genetic characteristics and to respect their uniqueness and diversity (2b).

Both instruments contain a range of provisions that apply their basic normative emphasis in contexts entailing or relevant to medical use. A good example of this being done well is the UDBHR s approach to authorizing research. Unlike the CHRB, the UDBHR requires research on the incapable to be compatible with their best interests (Art. 7a). It also states that

> (i)n appropriate cases of research carried out on a group of persons or a community, additional agreement of the legal representatives of the group or community concerned may be sought. In no case should a collective community agreement or the consent of a community leader or other authority substitute for an individual's informed consent.
>
> (Art. 6(3))

This third generation rights emphasis is also built out in Art. 16 and 17 to a focus on the need for restraint in the life sciences having regard for the interconnectedness of human beings with future generations.

However, as with the CHRB, one finds contexts involving or relevant to medical use in which respect is not fully protected. The UDHGHR, like the CHRB, expresses a basic but incomplete opposition to commodification, stressing via Art. 4 that the human genome in its natural state should not give rise to financial gains but doing nothing to prevent windfall gains that occur from work that draws on it. Furthermore, whilst Art. 11 of the UDHGHR prohibits practices which are contrary to human dignity, it only took the opportunity to give reproductive cloning as a specific exemplar. The creation of clones in the first place would surely have been made a further exemplar but for the fact that a few states would probably have resisted that. Eight years later, the General Assembly of the United Nations approved UNESCOs 2005 Declaration on Human Cloning (59/280) which calls on member states to prohibit such cloning but only by a majority of 84–34 (with 37 abstaining) because many states wanted to leave open the possibility of creating clones for research purposes. Contrary to a strong identity influenced approach to dignity, the UDHGHR also allows research, treatment and diagnosis that affects an individual's genome, subject to certain constraints (see Art. 5).

Conclusion

Global general human rights instruments have little in the way of medical use specific provision and much as their general emphasis on human rights and dignity and specific protections of the body are helpful, their implications for medical use can only be broadly guesstimated rather than precisely defined in the relative absence of relevant jurisprudence. Whilst this should be regarded as an unsatisfactory situation, it has to some extent, had the opportunity to be ameliorated through the development of health specific global human rights instruments.

Beyleveld and Brownsword (2001) have suggested that there has been a value shift in the orientation of these health specific instruments over time. Specifically, they argue that seen in terms of dignity older instruments such as the Nuremberg Code (1947) and the Declaration of Helsinki (1964) place a strong emphasis on its freedom dimension and that newer ones such as the CHRB and UDHGHR place greater emphasis on its constraint. Attempting to exemplify this, they suggest that central to the Code 'is the idea that human beings should not be subjected to scientific and medical research without their free and informed consent' (2001, 29) and that

> (t)o the extent that human dignity has a role to play in such thinking it is as the foundation for human rights, specifically the right of human beings to decide whether or not they will subject themselves to medical trials or treatments.
>
> (2001, 29)

By contrast, they suggest that in the latter instruments dignity is beginning to be treated as something more:

> we suggest that a conception of human dignity as constraint is implicated in much recent thinking about the limits to be placed on biomedicine, reflecting the fact that biomedical practice in the twenty first century should be driven, not by the vagaries of individual choice, but by a shared vision of human dignity that reaches beyond individuals. In other words, if we think of respect for human dignity as one of the constitutive values of our society (whether as an element of the public interest, or of *ordre public*, or as one of the fundamental values of our community), then those individual preferences and choices that are out of line with respect for human dignity are simply off limits.
>
> (2001, 29)

However, this summary is misleading. The instruments it cites as examples of older health related human rights instruments being largely oriented toward freedom rather than constraint are not. The Nuremberg Code has a principle which emphasizes the need for informed consent to research on human subjects and a further principle emphasizing the freedom to withdraw such consent. However, its eight other principles are devoted to ensuring that (even willing) prospective subjects are not taken advantage of. Its overall approach toward dignity or worth is rounded. Broadly speaking, the same could be said of the Declaration of Helsinki. Its initial development, and subsequent iterations, may be criticized as an attempt to displace the Code with something that at times is equivocal in its protection of welfare (Annas, 2003, 177) but that equivocation has taken the form of an attempt to dilute protection of the individual to meet demands for research rather than the form of an attempt to assert an 'excessive' emphasis on autonomy.

Furthermore, more recent health focused human rights instruments do not evince any ideological shift towards constraint. Older health related human rights instruments were typically limited in their objectives and/or topic focus whereas newer ones act as an incomplete but still fairly wide-ranging template for regional and domestic provision in the health field and also have to cater to threats to dignity that simply did not exist in the past (i.e. have emerged in the wake of scientific and technological developments). That newer instruments tend to have a greater constraint-oriented coverage is thus hardly surprising. It does not evidence that they have ideologically moved in the direction of constraint. Meanwhile, there is evidence to suggest that they have actually moved in the opposite direction to reflect an ongoing social and legal shift toward greater emphasis on autonomy in the medical context as well as more generally. They tend to have strong protections of individual rights with respect to informed consent and related matters such as rights to information about health and privacy. They also tend to reflect the social evolution that has occurred in the direction of greater freedom of action, falling shy of demanding that now somewhat accepted but once near universally condemned practices are prohibited. Notably, for example, there is no Convention or Declaration provision which stipulates that assisted suicide and so-called mercy killing should be prohibited.

In sum, global level provision for human rights both generally and in the health context has not so much ideologically shifted on its axis over the course of the post-World War II period as evolved itself principally to cater for the changing nature of actual and potential threats to respect for human beings. Mid-twentieth century instruments were heavily concerned with protecting the human rights of the individual vis-à-vis the state in the wake of many millions of people having suffered and died at the hands of fascist, expansionist powers. In the health field, they were not surprisingly concerned with the fact that atrocities carried out under fascist and even some non-fascist regimes were not infrequently at least partly medical in character. New concerns, not least those stemming from the rapidly emergent field of biotechnology, have invited new governance and a more explicit and elaborate focus on dignity and its constituent and related values.

Another point to make is that, whilst health specific human rights instruments are becoming more explicit and detailed in their focus on dignity and its relationships, they have considerable room for further improvement in this respect and with regard to their value orientation both generally and in the specific context of medical use. All of them, for example, could better articulate the core norms relevant to health and their relationships. The most evolved in this regard is the CHRB, but it could better define the scope of dignity as elaborated on in Art. 1 and its relationships with other norms that are focused on in this article, i.e. identity, equality and integrity. It could also explore the relationship dignity has to norms which are explicit or implicit elsewhere – such as autonomy, protection of well-being and non-commodification. This would not only be of value generally but aid the more specific task of defining constraints in the context of medical use. To the end of promoting such constraint, Art. 2 of the Convention could also be reformed to more explicitly link its emphasis on the primacy principle with an agenda of protecting human beings as ends and more explicitly link this agenda with protecting dignity. It ought, furthermore, to unequivocally state that both are absolute and be reformed such that its provisions are aligned to that.

3.3. Regional initiatives

European initiatives

The most important of these is the European Convention on Human Rights 1950 (ECHR). It has a significant range of substantive protections pertinent to protecting the body related interests of individuals and thus to constraining medical use of them. The most generically relevant is the Art. 8(1) right to private life. It protects autonomy (*Pretty v the United Kingdom* (2002) App no 2346/02 para. 61; *R.R. v Poland* (2011) App no 27617/04, para. 180) and physical and moral integrity (*YF v Turkey App no 24209/94* (2004) 39 EHRR 3). As a result performing any form of medical intervention engages it (*X v Austria App no 8278/78* (1979) 8 DR 154) as do the related practices of obtaining, retaining and using medical data (*S and Marper v UK* App no 30562/04 and App no 30566/04 [2008] ECHR 1581). The Art. 8(1) right to family life is also relevant in some contexts, most obviously in relation to an individual who has died or is living but does not have the legal power to decide the matter in question, such as for reasons of age or incapacity.

To avoid constituting violations of Art. 8, state restrictions on Art. 8(1) rights must meet the two requirements laid out in Art 8(2). *Firstly*, they must be in accordance with the law. Case law has established that this means that they must be based in national law with the law in question being accessible to the person concerned, foreseeable in its consequences for those subject to it and compatible with the rule of law (*Herczegfalvy v Austria* App no 10533/83 (1992) 15 EHRR 437. See also *McLeod v United Kingdom*, Case 24755/94, Judgment 23 September 1998 and *Hashman and Harrup v United Kingdom*, Case 25594/94, Judgment 25 November 1999). *Secondly*, they must be necessary in a democratic society for the protection of at least one of the legitimate aims laid out in Art. 8(2) (namely, national security, public safety, economic well-being of the country, prevention of disorder or crime, protection of health, protection of morals and protection of the rights and freedoms of others).

Meeting this requirement entails showing that the interference answers a 'pressing social need' and for the national authorities to find reasons for it that are relevant and sufficient having regard to its nature and extent (*S and Marper v UK* [2008] ECHR 1581, para. 101), albeit they are afforded some margin of appreciation in this assessment. Restrictions predicated on the conflicting rights of others have sometimes satisfied this criterion. However, those predicated on the mere needs or desires of others have not. For example, taking saliva and blood samples from criminal suspects for criminal justice purposes can satisfy it in certain circumstances (*X v the Netherlands* App no 8239/78 (Commission decision, 4 December 1978) Decisions and Reports (DR) 16, 187–189) and *Schmidt v Germany* App no 32352/02 (2006)), but blanket retention of body material samples and data from persons not convicted after subjection to a criminal process will not (*S and Marper v UK* App nos 30562/04 and 30566/04 [2008] ECHR 1581). Similarly, compulsory screening for tuberculosis during a serious outbreak can satisfy it in certain circumstances (*Acmanne v Belgium* (1984) 40 D.R. 251), but non-consensual gynaecological examination of a detained suspect on the dubiously claimed basis of protecting the police against allegations of sexual assault will not (*Juhnke v Turkey* App no 52515/99 Unreported May 13, 2008 (ECHR), para. 81).

Performing medical intervention (and thus medical use of the body) can also affect the Art. 9(1) freedom of thought, conscience and religion but not usually in a way that necessitates its separate consideration to the Art. 8(1) right to private life (*Hoffmann v Austria* (1993) Series A, No 255; *Pretty v the United Kingdom* (App no 2346/02)). It can also contribute to a treatment of a person that is severe enough to reach the threshold of being inhuman and degrading for the purposes of Art. 3. Whether it does so is considered in the light of all the circumstances, including its: duration; physical and mental effects; (in some cases) the gender, age and state of health of the victim; the mental state behind its infliction; *and* context, such as whether there was an atmosphere of heightened tension and emotions (see, for example, *M.S. v Croatia (No 2)* App no 75450/12 (2015). An example of its violation is *Jalloh v Germany* App no 54810/00 15 HRCD (2004–5) where a suspected drug trafficker who had been seen swallowing a suspected drug package was administered an emetic in circumstances where the less intrusive and less risky method of seizing the package after it had naturally passed through the suspects system could have been adopted. Art. 2 is relevant in selective cases where medical use threatens life, with the state having a positive duty to safeguard the lives of those in its jurisdiction.

Through Art. 14, the enjoyment of substantive ECHR rights and freedoms is protected from discrimination on any ground. Only with objective and reasonable justification can persons be treated differently in relation to them when they are in relevantly similar positions or the same in relation to them when they are in relevantly dissimilar positions (*Abdulaziz, Cabales and Balkandali v the United Kingdom*, judgment of 28 May 1985, Series A no 94, para. 71, and *Karlheinz Schmidt v Germany*, judgment of 18 July 1994, Series A no 291-B, para. 22). Furthermore, through Art. 13, there is a right to an effective remedy before a national authority in the event of violation of one or more of these substantive rights and freedoms. In the event of doubt over whether this right has been protected, there is the possibility of recourse to the European Court of Human Rights which has the power to deliver an adverse ruling on the member state concerned. At the same time as retaining ultimate sovereignty over their law, member states tend not only to respond appropriately to adverse rulings but also structure their domestic law and practice to minimize risk of them arising in the first place.

Backed by a reasonably interventionist Court that draws on CHRB provisions where relevant, the ECHR should provide considerable protection in the context of medical use. However, it does have weaknesses in this context. One is that individuals who have freely and capably agreed to be subject to conduct inconsistent with their worth are unlikely to draw on it. Their protection would instead rely on legal restrictions placed on such conduct at the domestic level – for example, provisions rendering purported consent to certain interventions and/or more generally interventions that are inconsistent with dignity/public policy invalid. Like many other human rights instruments, the ECHR does at least have a record of being interpreted in such a way as to deny the claims of those who assert they have a right to be free to be subject to such conduct (see, for example, *Laskey, Jaggard and Brown v UK* (1997) 24 EHRR 39). A second weakness with the ECHR is that like many other instruments, it does not fully protect all members of the human species. The statement within Art. 1 of the ECHR that parties to it have the obligation to secure to everyone the rights and freedoms contained within the Convention has not been interpreted as covering the not yet born or the dead. The Court does accept that both have interests, but only accounts for these in the context of domestic authorities using them to help defend conduct which would otherwise be viewed as incompatible with rights. A good example, is the case of *Poluhas Dödsbo v Sweden* App no 61564/00 (ECtHR, 17 January 2006) where the Court treated a dead man's interest in being left alone as a basis for concluding that the actions of the Swedish authorities in refusing a request by his grandchildren to move his ashes from a graveyard in Fagersta to one in Stockholm, where they lived, were necessary in a democratic society. Otherwise, the dead are reliant on their interests being incidentally protected when they overlap with the interests of next of kin (see further Garwood-Gowers & Pereira, 2017, 415–416).

An inclusive dignity-based position in relation to embryos would result in them being free of harmful medical use. The ability of a European jurisdiction to vindicate such freedom is somewhat circumscribed by the way the ECtHR has interpreted the 'right' to reproduce. In *Costa and Pavan v Italy* (ECtHR, 28th August 2012), the Court treated a prohibition on pre-implantation genetic diagnosis in Italy's law to be a disproportionate interference with the Art. 8(1) right to private life rights of a couple who wanted to use it to select between in-vitro embryos in order to identify one that was not carrying cystic fibrosis. Italian law was viewed by the Court to be morally inconsistent on the basis that it would have been legal for the woman of the couple to have had an abortion had she been pregnant with a foetus that was carrying cystic fibrosis. This view, however, does not account for arguments that pregnant women should be seen as having an extensive or even absolute right not to continue carrying an embryo/foetus based on the right to bodily security (Garwood-Gowers, 2008). Ironically, the Court has been more willing to allow states to unjustifiably restrict this right in order to protect embryos (*ABC v Ireland*, App no. 25579/05, 16 December 2010) than to justifiably restrict choices about assistance in reproduction in order to protect embryos. It would also seem that whilst the bodily security of women is not being properly validated, there is a kind of neo-liberal freedom to exploit embryos both for the reproductive business and more generally. It may well be that the European Charter of Fundamental Rights 2000 ends up being a better tool for curbing this freedom than the ECHR. Art. 1 of it emphasizes the inviolability of dignity. In *Oliver Brüstle v Greenpeace eV C-34/10*, it was used to support a conclusion that the EU Directive on the legal protection of biotechnological inventions (98/44/EC) does not extend to allowing patenting of stem cells derived from a human embryo. The Charter also contains some specific biotechnology related provisions that are designed to protect dignity in a manner that goes beyond the individual born person – in particular Art. 3(2) prohibitions on eugenic practices (in particular those aiming at the selection of persons), commodification via the human body and reproductive cloning of human beings.

Inter-American initiatives

The creation of supra-national human rights instruments in the Americas started with the several hundred years before the European invasion with the Haudenosaunee Great Binding Law of Peace. The first of pan-American scale was the so-called Bogotá Declaration – the American Declaration of the Rights and Duties of Man 1948 – which as the first supra-national human rights instrument to rise from the ashes of the World War II not surprisingly includes a rebuttal of totalitarian thought with its emphasis on the essential rights of human beings deriving from attributes of human personality as opposed to nationality. This idea is then linked in the preamble to a set of more profound notions including that humans should serve spiritual development as the supreme end of human existence and the highest expression thereof; are born free and equal, in dignity and in rights; and should act with solidarity.

Such sentiments tie in quite nicely with assertions about human nature and purpose I made in Chapter 1, but it is potentially problematic that the Declaration grounds duty based substantive provisions on them, including: Art. 29, which echoing Kant's duty of perfection requires the individual 'so to conduct himself in relation to others that each and every one may fully form and develop his personality,' and Art. 34, which imposes a somewhat more tangible and specific duty on 'every able-bodied person to render whatever civil and military service his country may require for its defence and preservation, and, in the case of public disaster, to render such services as may be in his power.' However, the concern that this could be used by states to justify coerced medical use is mitigated by three factors. Firstly, the duty appears to be based around an obligation of mutual defence – in the form of doing what is strictly necessary to meet military and natural threats which could affect any or all of those subject to the duty. It is only very selectively the case that medical use would be a pertinent tool for meeting such threats. Secondly, it is one thing to expect people to give up their labour and/or put themselves at bodily risk in such circumstances but arguably much more intrusive to expect them to forcibly subject to medical intervention. Thirdly, the Declaration protects a range of rights which might mitigate against such a possibility – such as the right to life, liberty and security of the person (Art. 1) and the right to private and family life (Art. 5).

In several respects that are significant in the context of medical use, the Declaration has not only been built on but superseded in significance by the American Convention on Human Rights 1969. This has a broader range of civil and political rights. The most pertinent of these for constraint of medical are its rights to life (Art. 4), humane treatment (Art. 5), freedom from slavery (Art. 6), personal liberty (Art. 7), freedom of movement (part of Art. 22), privacy (Art. 11), family life (Art. 17), freedom of conscience and religion (Art. 12) and freedom of thought (part of Art. 13). Like the ECHR, it emphasizes the need for the rights and freedoms it protects to be protected without discrimination on any ground (Art. 1). The Convention is also less threatening in its approach to duties than the Declaration. Its statement that everyone 'has responsibilities to his family, community and mankind' (Art. 32(1)) is too imprecise in character to be treated as a basis to restrict right and its emphasis on rights being limited by 'the rights of others, by the security of all, and by the just demands of the general welfare, in a democratic society' (Art. 32(2)) is qualified by the need for democracy and may, therefore, preclude the possibility of instrumentalizing medical use. Importantly, the Convention also represents the culmination of efforts to set up an Inter-American Commission on Human Rights and an Inter-American Court for the Protection of Human Rights (Brownlie & Goodwin-Gill, 2006, 933). The result is an ECHR style framework of enforcement of rights and freedoms: The Commission can receive petitions, denunciations and complaints of violation of the Convention by a State Party from any person or group of persons as well as nongovernmental entities legally recognized in one or more member States of the Organization of American States (Art. 44), *and*, in the absence of settlement, there are procedures for the Court to accept jurisdiction upon submission to it by the Commission or a concerned State party (Art. 51) and injured parties can obtain remedies for violations (Art. 63).

There are also a range of specific-issue Inter-American conventions that are relevant in the medical use context. Perhaps the most significant is the Inter-American Convention on the Elimination of all Forms of Discrimination Against Persons with Disabilities 1999 which treats the inherent dignity and equality of each person as its grounding and contains a range of measures to achieve its objective of preventing and eliminating disability discrimination and promoting full integration.

African initiatives

The most important African human rights instrument is the African Charter on Human and Peoples Rights 1981, commonly known as the Banjul Charter. Compliance with the rights protected by the Charter is supported through the African

Commission on Human and People's Rights (set up under Art. 30) and, since 2004, a Court. The Charter includes a normative emphasis on dignity within its preamble which Art. 5 carries through into an emphasis on everyone having the right to the respect of the dignity inherent in a human being and to the recognition of his legal status and prohibits all forms of exploitation and degradation of man (particularly slavery, slave trade, torture, cruel, inhuman and degrading punishment and treatment). The Charter additionally emphasizes equality in a number of ways. For example:

- Art. 2 stresses that 'everyone is entitled to the enjoyment of rights and freedoms recognized and guaranteed in the Charter without distinction of any kind';
- Art. 3 emphasizes equality before the law and equal protection of the law;
- Art. 19 stresses equality at the population level – which mean not only at the level of people within a state, but also at the level of a group within a state who see themselves as distinct by reason of common ancestry, ethnic origin, language or cultural habit and also at the level of indigenous peoples (see further, Baldwin & Morel, 2008, 247);
- Art. 28 places on every individual a duty 'to respect and consider his fellow human beings without discrimination, and to maintain relations aimed at promoting, safeguarding and reinforcing mutual respect and tolerance'; *and*
- Art. 18 requires states to take action to 'ensure elimination of discrimination against women and ensure the rights of women and children are protected as stipulated in international declarations and conventions.' The rights of these groups are also specifically respectively supported by the Charter's Protocol on the Rights of Women in Africa 2003 and the African Charter on the Rights and Welfare of the Child 1990, both of which contain a range of rights pertinent to protection of body related interests.

Dignity and equality have additionally been made more tangible in the health context through a range of Commission decisions and other institutional initiatives. Perhaps the most relevant in the immediate context is the case of *Purohit and Moore v The Gambia* (Communication 241/2001) in which individuals compulsorily detained on mental health grounds, mostly poor people picked up off the streets, were found to lack an accessible remedy. In the same case the Commission interestingly used Art. 5 to stress that persons with, or treated as having, mental illness had to be treated with humanity and respect for the inherent dignity of the human person and like other human beings had 'a right to enjoy a decent life, as normal and full as possible, a right which lies at the heart of the right to human dignity' (para. 61). With the exception of the dubious reference to the word 'normal,' one can appreciate these sentiments as well as the substance of the conclusion that it was contrary to dignity to warehouse such a person in poor and overcrowded conditions and label them lunatics and idiots.

The Charter also contains rights and protections focused on the body and thus relevant to its medical use. Most notably, Art. 4 stresses that human beings are inviolable and that every person is entitled to respect for his life and the integrity of his person. It not only protects against actual intrusions on these but threats of them (Communication 205/97). Art. 29 of the Charter places a duty upon the individual to 'serve his national community by placing his physical and intellectual abilities at its service.' As with similar provisions in Articles 29 and 34 of the American Declaration, this appears to generate a duty to rescue that will crimp freedom from forced bodily action but is unlikely to extend to allowing instrumentalizing medical use particularly as the Charter, like the Declaration, protects dignity and various body related rights.

Overall the substance of the Charter may be commended in the medical use context and more generally. This substance has been defended in two crucial ways. Firstly, unlike its European and inter-American counterparts, the Charter does not permit derogation. Secondly, the fact that the Commission – despite weakness in its independence and resourcing (see Baricako, 2008, 9 and 12–13) has prevented rights from being interpreted by member states in a way that render them largely or wholly ineffectual. For example, in a number of decisions, it has stated that international human rights standards must prevail over contradictory national law (Communications 105/93, 128/94, 130/94 and 152/96) with the former requiring restrictions on rights be not only lawful but necessary. The main weaknesses of the system lie in the fact that application to the Court by NGOs and individuals only occurs in relation to member states who have made a declaration permitting it and the fairly weak overall track record of states in effectively implementing measures in response to findings against them (Manby, 2008). Even given this, the Charter has proved to have considerably more bite than the kind of paper tiger that critics such as Umozurike predicted it would be (1988, 82–83).

Arab initiatives

Two Arab human rights instruments are significant in the context of constraint of medical use. *Firstly*, the Cairo Declaration on Human Rights in Islam (1990). This emphasizes dignity and non-discrimination (Art. 1, 6 and 19). It also contains a number of substantive protections relevant to the body including: The right to life (Art 2); freedom from slavery (including colonialism), humiliation, exploitation and oppression (Art. 11); rights to security and privacy (Art. 18); *and* freedom from physical or psychological torture, humiliation, cruelty and indignity and from medical or scientific experimentation without consent or at risk to health or life (Art. 20). *Secondly*, the Arab Charter on Human Rights 2004 which also contains various substantive protections relevant to the body, including: the right to not be arbitrarily deprived of life (Art. 5); freedom from physical or psychological torture *and* cruel degrading, humiliating and inhuman treatment (Art. 8); freedom from slavery, forced labour, exploitation and trafficking in human beings (Art. 10); freedom of thought, conscience and religion (Art. 30); and freedom from medical or scientific experimentation and use of organs without free informed consent and due regard to personal safety (Art. 20). The

Charter also stresses non-discrimination in enjoyment of the rights and freedoms set out in it (Art. 3) and contains an undertaking on the part of States to 'ensure to persons with mental or physical disabilities a decent life that guarantees their dignity, and to enhance their self reliance and facilitate their active participation in society' (Art. 40). The main systematic problem with the Arab approach to protecting human rights is the weakness of its enforcement system. Pursuant to Art. 45 of the Charter, an Arab Human Rights Committee has been established, but the Committee's functions appear to be limited to considering and producing reports. As yet it lacks a regional Court with which to enforce rights granted under the Charter.

3.4. Exemplars of protections of body related rights in the domestic sphere

US public law

The US has a system which individual human rights legal primacy through the fact that the Bill of Rights 1789 has been at the apex of its Constitution since the Supreme Court determined in *Marbury v Madison* 5 US (1 Cranch) 137 (1803) that provisions of state and national law that conflicted with it could be voided. Deviations from protection of rights of the individual are carefully circumscribed under the Bill. The Fourteenth Amendment right not to be deprived of life, liberty of property without due process of law is particularly relevant in the context of medical use as illustrated by the case of *In re Cincinnati Radiation Litigation* 874 F Supp 796 (SD Ohio 1995). Here, the US District Court for the Southern District of Ohio, Western Division, allowed the heirs and personal representatives of now deceased cancer patients to proceed on constitutional grounds against government and university physicians who used them in radiation exposure experiments without any or proper informed consent. The background to this was that, from 1960 to 1972, the Department of Defense funded research on radiation exposure to humans, which was administered by the University of Cincinnati College of Medicine and performed at Cincinnati General Hospital. Their research subjects were cancer patients with life expectancies of 2 years or less, usually indigent, poorly educated and of lower than average intelligence. The majority of them were African-American. The patients were lied to by being told that the radiation was for the benefit of treating them and not told that the radiation dosage was determined by the experimental design of the research, 'to study the effects of massive doses of radiation on human beings in preparation for a possible nuclear war.' No consent forms were used for the first 5 years. When consent forms came into use, the risk of radiation exposure to the patients was not stated. The court found that the defendants were not acting as physicians, but instead were acting 'as scientists interested in nothing more than assembling cold data.' Consequently, their actions were deemed to be outside the scope of their hired duties to care for the sick and injured and thus not immune to prosecution under the Due Process Clause of the Fourteenth Amendment for violations of liberty and equal protection of rights.

UK public law

By contrast to the US, the UK system has traditionally been seen as based on the idea that Parliament is the ultimate authority as reflected in the doctrine of Parliamentary sovereignty. However, the nature and extent of Parliament's authority have been drawn into question over time, not least through such trends as concentration of power in the executive, devolution, Europeanization and the introduction of the Human Rights Act (1998). The latter is especially important in the immediate context because it has given Courts and tribunals a legal duty to act compatibly with nearly all ECHR rights as far as it is possible to do so without acting incompatibly with the will of Parliament. When it is not possible to do so, they can issue a section 4 declaration of incompatibility. In certain circumstances this will pave the way for the shortfall in protection to be removed by way of ministerial action under section 10 of the Act. It is notable that the duty to act compatibly with Convention rights applies not just to the Courts and tribunals (section 6(3)a) but also to 'any person certain of whose functions are functions of a public nature' (section 6(3)b).

Common law civil and criminal norms

Common law systems have criminal law provisions designed to protect the person that ranges from the likes of common assault through to the likes of grievous bodily harm and right up to murder. Civil law protection centres on the law of tort which renders all forms of intentional touching beyond the 'everyday' prima facie actionable as trespass to the person. Whilst consent is an important defence in both the criminal and civil context, it is a normal feature of common law systems to restrict what can be validly consented to as a matter of public policy. Medical intervention specifically has to meet the standard of being reasonable or proper for consent to it to be valid (see, respectively, *Attorney General's Reference (No 6 of 1980)* [1981] QB 715 and *R v Brown* [1994] 1 AC 212). What this means in the medical use context has never been fully articulated in jurisprudence. In the case of *Urbanski v Patel* (1978) 84 D.L.R. (3d) 650, it was stated *obiter dicta* that a medical professional could not rely on consent as a justification for taking a single remaining kidney from an individual for the purposes of living donor organ transplantation. This suggests that there is a threshold of prospective detriment (harm and/or risk) that cannot be exceeded and/or a requirement that prospective harm and/or risk to the person undergoing the intervention not be disproportionate to the prospective benefits obtainable from its performance (in the medical use context these would principally be benefits to another or the community at large). Equally, there is authority for the proposition that consent to non-therapeutic research must be founded on disclosure of all significant information (*Halushka v University of Saskatchewan* (1965) 52 WWR 608), and it may be suggested that the common law will require consent to any intrusion which deviates from the optimum protection of its subject's clinical interests to be predicated on disclosure of the fact, nature, extent and implications of that deviation.

These constraints, along with the protection the law supplies to those who lack sufficient functional capacities to give valid consent, illustrate how general law is driven by respect for human beings in the bodily context which on the one hand leaves those capable of making their own decisions free to do so to a large degree but on the other provides them, along with the incapable, with protection from exploitation. This approach appears to generate significant constraint in the context of medical use, but its full applicability in this context has sometimes been questioned. Skegg, for example, once wondered whether the courts might at least interpret necessity in such a way as to sanction very minimal bodily intrusions on the body of the individual to meet important needs:

> (i)t is just possible to envisage cases in which a court might grant an absolute discharge if criminal proceedings were brought. One such case might be where a patient with a rare blood group will die if his blood is not 'topped up', and the only way of obtaining the blood is to remove it from someone who is unable to consent.
>
> (1988, 17)

However, this proposition has no authority to support it. Furthermore, there are authorities to support the counter-argument that the common law defends the primacy of the individual in the context of efforts to dilute protection of his or her body related rights to meet the mere needs of others. The classic case in this regard is *McFall v Shimp* (1978) No. 78–17711. 10 Pa D & C (3d) 90 (Ct Comm Pl, Pa). Shimp had undertaken tests which had established he was genetically matched for the purposes of donating bone marrow to his cousin, McFall, an aplastic anaemia sufferer. When he decided not to undertake remaining tests to determine his clinical suitability for donation, McFall petitioned the Alleghany County Court to force him to do so and then to donate if he was found suitable. Rejecting the application for the Court, Mr Justice Flaherty emphasized that,

> The common law has consistently held to a rule which provides that one human being is under no legal compulsion to give aid or to take action to save another human being or to rescue. A great deal has been written regarding this rule, which, on the surface, appears to be revolting in a moral sense. Introspection, however, will demonstrate that the rule is founded on the very essence of our free society. Our society, contrary to many others, has as its first principle, the respect for the individual, and that society and government exist to protect the individual from being invaded and hurt by another. . . . For our law to compel the defendant to submit to an intrusion of his body would change every concept and principle upon which our

> society is founded. To do so would defeat the sanctity of the individual, and would impose a rule which would know no limits, and one could not imagine where the line would be drawn. . . . For a society, which respects the rights of *one* individual, to sink its teeth into the jugular vein or neck of one of its members and suck its sustenance for another member, is revolting to our hard-wrought concepts of jurisprudence. Forcible extraction of living body tissue causes revulsion to the judicial mind. Such would raise the spectre of the swastika and the Inquisition, reminiscent of the horrors this portends.
>
> (p. 91)

The common law has also insisted that those who are legally having decisions made for them (for example, some incapable adults and minors) have their interests protected even in the face of dilution of them having the potential to benefit another or society at large (see, for example, *Grimes v Kennedy Krieger Institute* (2001) 782 A.2d 807). Some common law and many non-common law jurisdictions do impose a duty to rescue in the accident, common danger and emergency situation. For example, Art. 64(2) of the French Criminal Code Art. 63(2) creates an offence of 'non-assistance à personne en danger' under which deliberately failing to provide assistance to a person in danger can be punished by up to 5 years of jail time and a fine of up to €100,000. Similarly, Art. 323c of the German Criminal Code creates an offence of neglect of duty to provide assistance ('Unterlassene Hilfeleistung') which includes an obligation, in most circumstances, to provide first aid when necessary (in Germany, knowledge of basic emergency measures is a prerequisite for the granting of a driving licence). However, whilst duty to rescue laws clearly interfere with bodily freedom and rights which related to that related to the body freedom of bodily action. To found bodily intrusion on the needs of others is a whole other level of interference with rights, though as we shall see in subsequent chapters there are some very specific legislative examples of support for it.

3.5. Conclusions

One of the questions that arises from this chapter is whether a single principle or a collection of principles can and/or should be regarded as the foundation for constraint of medical use of the human being. This ties into the question of what principle or principles are and/or should be treated as the foundation of the overlapping fields of bio law, medical law and health law in general? Efforts to answer this question have often suffered from being quite apparently incomplete but in general terms can be said to be evolving. Understanding of Dutch health law is a good example of this. It often used to be regarded as based on the right to self-determination, the right to health or a combination of these two principles (see further Dute, 2005; Buijsen, 2017) when in fact like health law in other jurisdictions it has always been concerned with protecting worth in a more all-round sense – not least, for example, given the role that protecting those who lack competence and capacity and protecting all from abuse and exploitation play in it. Many health lawyers have implicitly recognized this fact through treating dignity as the basis of health law (for example, Van der Burg, 1996; Brownsword & Bylefeld, 2001; Arends, 2004; Foster, 2011; Andorno, 2013; Buijsen, 2017) – an approach that as Buijsen notes (2017, 4), Dutch health lawyers have been willing to accept in the wake of its explicit recognition in Dutch Government Policy which, in turn, is influenced by the international standards to that effect already alluded to within this chapter.

Organic examination of what governance relating to health is and/or should be achieving can help us better judge between such competing approaches. Two goals are key: Firstly, obtaining or maintaining health and healthcare, to wit one can argue that the right to health is certainly *a* normative pillar of healthcare law; and, secondly, ethical constraints on medical practices, to wit the principle of self-determination is *but one of many* potentially relevant considerations. Both pillars can be related to a conception of respect for human beings that is grounded on dignity in its all-round protection of worth sense. However, restraint of medical use of the human being has very little to do with the first pillar and is a specialized function of the second pillar. Hence, the conclusions drawn in chapter on the implications of protecting worth in the context of medical use (treating all members of the species as of equal inherent, protecting human nature and form, avoiding instrumental use of human beings and appropriately weighting other relevant principles, such as autonomy and beneficence) can only be treated as definitive in this context, though they would also function as a useful starting point to understanding how to protect the second pillar in general.

This chapter has shown that generic and health specific human rights instruments tend to be structured in a manner that is broadly compatible with this understanding and are often specifically well suited to protecting against unjustifiable medical use. That is important not just in itself but also because some of the shortfalls of their specific provisions may be capable of being 'read down' in the light of their general normative orientation. For example, provisions appearing to permit some medical research on incapable adults and incompetent minors that would be inconsistent with their best interests could be read down in the light of provision emphasizing their dignity and/or primacy. However, some shortfalls – such as the fact that these instruments do not tend to apply fully to the not yet born and the dead – are difficult to read down.

References

Andorno, R. (2013) *Principles of International Biolaw: Seeking Common Ground at the Intersection of Bioethics and Human Rights*. Geneva: Bruylant.

Andorno, R. (2016) Is vulnerability the foundation of human rights? In Masserer, A. & Garcia-Sánchez, E. (eds.) *Human Dignity of the Vulnerable in the Age of Rights*. Heidelberg: Springer, pp. 257–272.

Annas, G. (2003) Questing for grails: Duplicity, betrayal and self-deception. In Tomossy, G. & Weisstub, D. (eds.) *Postmodern Medical Research, in Human Experimentation and Research*. Ashgate: Dartmouth, pp. 171–198.

Arends, L. (2004) *The Psychogeriatric Patient and the Law, Taking Care of Restrictions on Personal Liberty*. The Hague: Sduuitgevers.

Baldwin, C. & Morel, C. (2008) Group rights. In Evans, M. & Murray, R. (eds.) *The African Charter on Human and Peoples' Rights*, 2nd Edition. Cambridge: Cambridge University Press, pp. 244–288.

Baricako, G. (2008) Introductory preface: The African charter and African commission on human and people's rights. In Evans, M. & Murray, R. (eds.) *The African Charter on Human and Peoples' Rights*, 2nd Edition. Cambridge: Cambridge University Press, pp. 1–19.

Beyleveld, D. & Brownsword, R. (2001) *Human Dignity in Bioethics and Biolaw*. Oxford: Oxford University Press.

Brownlie, S. & Goodwin-Gill, G. (2006) *Basic Documents on Human Rights*, 5th Edition. Oxford: Oxford University Press.

Buijsen, M. (2017) Human dignity and the concept of health law. In den Exter, A. (ed.) *European Health Law*. Leiden: Brill, pp. 3–24.

Council of Europe, Directorate of Legal Affairs. (1997) *Explanatory Report to the Convention for the Protection of Human Rights and Dignity of the Human Being with Regard to the Application of Biology and Medicine*. Strasbourg, January, Doc. DIR/JUR (97)1.

Dute, J. (2005) The leading principles of the convention on human rights and biomedicine. In Gevers, J., et al., (eds.) *Health Law, Human Rights and the Biomedicine Convention*. Leiden; Boston, MA: Martinus Nijhoff, pp. 3–12.

Foster, C. (2011) *Human Dignity in Bioethics and Biolaw*. Oxford: Hart.

Gadd, E. (2001) General provisions of the convention. *Journal International de Bioethique*. 12(1), 21–49.

Garwood-Gowers, A. (2008) The right to bodily security vis-à-vis the needs of others. In Weisstub, D & Diaz-Pintos, G (eds.) *Autonomy and Human Rights in Healthcare: An International Perspective*. London: Springer, pp. 375–397.

Garwood-Gowers, A. & Pereira, A. (2017) Human to human organ and tissue transplantation. In den Exter, A. (ed.) *European Health Law*. Leiden: Brill, pp. 391–418.

Kayess, R. & French, P. (2008) Out of darkness into light? Introducing the Convention on the Rights of Persons with Disabilities. *Human Rights Law Review*. 1–34.

Loenen, T. (2001) Towards a common standard of achievement? Developments in International Equality Law. *Acta Juridica*. 197.

Manby, B. (2008) Civil and political rights in the African Charter on Human and Peoples' Rights: Articles 1–7. In Evans, M. & Murray. J. (eds.) *The African Charter on Human and Peoples' Rights*, 2nd Edition. Cambridge: Cambridge University Press, pp. 171–212.

Skegg, P. (1988) *Law, Ethics and Medicine*. Oxford: Clarendon Press.

Umozurike, U. (1988) The protection of human rights under the Banjul (African) Charter on Human Rights and Peoples' Rights. *African Journal of International Law*. 65, 82–83.

Van der Burg, W. (1996) Gezondheidsrecht en Bio-ethiek; Op weg naar een nieuwe verhouding [Health law and Bioethics: In search of a new relationship]. *R & R*. 193–206.

Part II

Shortfalls in constraint – three areas viewed through the lens of respect

4 Abusive warfare related medical use

4.1. Introduction

The prospect and actuality of war can lead to the amplification of many good and bad human tendencies. Amongst the latter is the tendency to use medical knowledge, technology and intervention in ways that damage human beings. This chapter considers two contexts in which such damaging use takes place: *firstly*, warfare related programmes which consist of inventing/discovering, producing, testing and weaponizing (one or more of) chemical, biological and radiological agents to the end of destroying human beings and/or the environs which they rely on, and *secondly*, abusive experimentation performed on human beings for both the aforementioned and other warfare related ends.

One of the ways in which this chapter adds to the discourse concerning such damaging forms of medical use is by encompassing the whole span of industrial-scale reliance on them: starting with the dawn of industrial-scale use of chemical warfare in World War I, going on to examine the use and long term societal impacts of various forms of warfare related experimentation in and around World War II and the cold war period and ending up briefly touching on the continuing problem of warfare related use in more recent times. The literature related to warfare related medical use is known for 'honing in' by reference to time frame, geographical or national context and type of use. By contrast I examine it in a long range and global fashion with a view to more holistically understanding when, how and why it occurs. I tease out what it has tell us, both about medical use and about what happens when philosophies that entail sacrificing respect are given their head.

4.2. World War I and the interwar years

The use of biological and chemical agents in warfare has been occurring on a regular basis for thousands of years (Guillemin, 2005, 3; Riedel, 2004, 400). Such agents have been used as a means of directly harming and killing individuals (such as through assassination with poison) and large groups (such as via targeting disease infected materials at a village, town or army) and have also in some cases been used to target the environments that they rely on (such as using chemicals, bacteria and viruses to target crops and water sources). During the course of the modern era, concerns about the threat posed by such unorthodox methods of warfare has amplified not least because the speed at which we have managed to develop relevant scientific knowledge and technology has not been matched by advances in our ability to respect our worth.

Reflecting the seriousness of the threat, the Hague Declaration (1899) included within it a pledge that states would 'abstain from the use of projectiles the object of which is the diffusion of asphyxiating or deleterious gases.' Having been one of the many parties to ratify the Declaration, the German empire went on to catastrophically and systematically violate it over the course of World War I. This began with releasing chlorine gas towards the Allied trenches at Ypres on the 22nd April 1915. By mid-1916, both sides had started to use phosgene which was roughly 18 times as powerful as chlorine (Harris & Paxman, 1982, 18). The vagaries of wind speed and direction were then greatly reduced in 1916 and 1917 by the development of targeted means of delivery. The British developed tubular steel projectors out of which they fired drums of phosgene and other chemicals whilst the Germans used shells to convey chemicals, which by mid-1917 included mustard gas (dichlorethyl sulphide). In all, about 113,000 tonnes of chemicals were used as weapons in the war, wounding around 1.3 million men (Harris & Paxman, 2002, 32).

The scientific and technological developments that led to such extreme warfare were built partly on experimentation. For example, the British, having developed the world's first chemical warfare installation at Porton Down in 1916, soon began to conduct extensive harmful testing of agents on the environment, animals and human beings:

> A system of circular trenches was dug, from the centre of which cylinders of gas were discharged. Human guinea pigs ('observers' in Porton's terminology) would station themselves in trenches and – for as long as they were capable of standing it – take detailed notes of the symptoms they felt. Indoors, the effects of chemicals were studied in the gas chambers. Ten minutes was found to be about the maximum most men could take exposed to a non-lethal gas. Observers were expected to stand in clouds of killer gases for hours wearing prototype masks to test their reliability. Later, when mustard gas made its first appearance, they rolled up their sleeves and allowed their arms to be contaminated, in order to study the progression of the terrible blisters that developed.
> (Harris & Paxman, 1982, 39)

In 1917, a farm and breeding colony was added to the establishment and went on to provide vast numbers of animals for experimentation, including monkeys and baboons. At the same time, the demand for 'virgin' humans to test on far exceeded supply, such that cooks, orderlies and clerks were frequently pressed into service. It was obvious from an early stage that the tests were dangerous. Indeed, some volunteers died before the war was out.

Most major powers, treated the post-war period as an opportunity to stockpile and to develop further offensive measures. However, they did not always resist drawing on those stockpiles and new measures for their delivery when they ended up in conflicts. In 1919 British interventions in the Russian Civil War included supplying the White Army with gas shells and dropping arsenic canisters at Archangel

(Harris & Paxman, 2002, 43). Later that year, the British appear to have used phosgene and mustard gas on the Afghan frontier to flush out 'rebellious tribesmen' from caves (Harris & Paxman, 2002, 44). By 1925, the French and Spanish were employing poison gas to curb rebellion in Morocco, and in 1936 Italy treated its war in Abyssinia as a proving ground for the use of mustard gas bombs.

Inevitably, there was also a continued toll exacted through experimentation. The British, for example, continued to scientifically explore the effects of mustard gas and other chemical agents not just through studies of World War I victims (who had the effects of their exposure downplayed by the Government for the purpose of harshly limiting their war pensions), but also through testing their effects on thousands of volunteers (Harris & Paxman, 2002, 35 and 43). This, however, was nothing compared to the scale of atrocious general and unconventional warfare related experimentation conducted by the Japanese in the territories that they started to invade from the early 1930s and by the Germans on prisoners and inmates of concentration camps during World War II.

4.3. Germany's World War II experimental atrocities

The German World War II programme of experimental atrocities can be characterized as having two distinct aspects: firstly, volunteers were used to carry out tests on chemical agents. The chemical used included newly discovered deadly nerve agents, some of which were mass produced and although not used on the battlefront were used to exterminate millions of concentration camp detainees. Secondly, many concentration camp detainees and prisoners were used for a wide range of experiments (Harris & Paxman, 1982, 60–62). Revelations that affect our understanding of the nature, extent and extremity of such experiments continue to emerge (see, for example, Döring, 2014, 145–149). However, a good flavour of them had already been presented at the so-called Doctors Trial at the Nuremberg Military Tribunals (see Figure 4.1).

Three further sets of experiments not mentioned on the indictment were identified as having occurred during the Doctors Trial (Harvard Law School, 2018), namely:

- *Phenol (gas oedema) experiments* (1942–1944). Conducted to investigate whether levels of phenol in gas oedema serum caused fatalities among wounded soldiers; experiments were conducted on prisoners at Buchenwald.
- *Phlegmon experiments* (1942). Conducted to test treatments for sepsis and related diseases, in co-ordination with sulfanilamide experiments at Ravensbrueck; experiments were conducted at Dachau and Auschwitz.
- *Polygal experiments* (1943–1944). Conducted to test the effectiveness of polygal, a blood coagulant, for the treatment of wounds.

Some of the more general medical atrocities that had been perpetrated during the war were also described at the Trial, including euthanasia on parts of the German population, the imprisonment and (in certain cases) killing of Poles who had incurable tuberculosis and the killing of 112 Jews at Auschwitz for the

COUNT TWO – WAR CRIMES

(A) High Altitude Experiments. From about March 1942 to about August 1942, experiments were conducted at the Dachau concentration camp, for the benefit of the German Air Force, to investigate the limits of human endurance and existence at extremely high altitudes. The experiments were carried out in a low-pressure chamber in which atmospheric conditions and pressures prevailing at high altitude (up to 68,000 feet) could be duplicated. The experimental subjects were placed in the low-pressure chamber and thereafter the simulated altitude therein was raised. Many victims died as a result of these experiments and others suffered grave injury, torture, and ill-treatment.

(B) Freezing Experiments. From about August 1942 to about May 1943, experiments were conducted at the Dachau concentration camp, primarily for the benefit of the German Air Force, to investigate the most effective means of treating persons who had been severely chilled or frozen. In one series of experiments the subjects were forced to remain in a tank of ice water for periods up to 3 hours. Extreme rigor developed in a short time. Numerous victims died in the course of these experiments. After the survivors were severely chilled, re-warming was attempted by various means. In another series of experiments, the subjects were kept naked outdoors for many hours at temperatures below freezing. The victims screamed with pain as their bodies froze.

(C) Malaria Experiments. From about February 1942 to about April 1945, experiments were conducted at the Dachau concentration camp in order to investigate immunization for and treatment of malaria. Healthy concentration-camp inmates were infected by mosquitoes or by injections of extracts of the mucous glands of mosquitoes. After having contracted malaria, the subjects were treated with various drugs to test their relative efficacy. Over 1,000 involuntary subjects were used in these experiments. Many of the victims died and others suffered severe pain and permanent disability.

(D) Lost (Mustard) Gas Experiments. At various times between September 1939 and April 1945, experiments were Conducted at Sachsenhausen, Natzweiler, and other concentration camps for the benefit of the German Armed Forces to investigate the most effective treatment of wounds caused by Lost gas. Lost is a poison gas which is commonly known as mustard gas. Wounds deliberately inflicted on the subjects were infected with Lost. Some of the subjects died as a result of these experiments and others suffered intense pain and injury.

(E) Sulfanilamide Experiments. From about July 1942 to about September 1943, experiments to investigate the effectiveness of sulfanilamide

Figure 4.1 Medical experiments as war crimes at the Doctors Trial.

were conducted at the Ravensbrueck concentration camp for the benefit of the German Armed Forces. Wounds deliberately inflicted on the experimental subjects were infected with bacteria such as streptococcus, gas gangrene, and tetanus. Circulation of blood was interrupted by tying off blood vessels at both ends of the wound to create a condition similar to that of a battlefield wound. Infection was aggravated by forcing wood shavings and ground glass into the wounds. The infection was treated with sulfanilamide and other drugs to determine their effectiveness. Some subjects died as a result of these experiments and others suffered serious injury and intense agony.

(F) Bone, Muscle, and Nerve Regeneration and Bone Transplantation Experiments. From about September 1942 to about December 1943, experiments were conducted at the Ravensbrueck concentration camp, for the benefit of the German Armed Forces, to study bone, muscle, and nerve regeneration, and bone transplantation from one person to another. Sections of bones, muscles, and nerves were removed from the subjects. As a result of these operations, many victims suffered intense agony, mutilation, and permanent disability.

(G) Sea-water Experiments. From about July 1944 to about September 1944, experiments were conducted at the Dachau concentration camp, for the benefit of the German Air Force and Navy, to study various methods of making sea water drinkable. The subjects were deprived of all food and given only chemically processed sea water. Such experiments caused great pain and suffering and resulted in serious bodily injury to the victims.

(H) Epidemic Jaundice Experiments. From about June 1943 to about January 1945, experiments were conducted at the Sachsenhausen and Natzweiler concentration camps, for the benefit of the German Armed Forces, to investigate the causes of, and inoculations against, epidemic jaundice. Experimental subjects were deliberately infected with epidemic jaundice, some of whom died as a result, and others were caused great pain and suffering.

(I) Sterilization Experiments. From about March 1941 to about January 1945, sterilization experiments were conducted at the Auschwitz and Ravensbrueck concentration camps, and other places. The purpose of these experiments was to develop a method of sterilization which would be suitable for sterilizing millions of people with a minimum of time and effort. These experiments were conducted by means of X-ray, surgery, and various drugs. Thousands of victims were sterilized and thereby suffered great mental and physical anguish.

(J) Spotted Fever (Fleckfieber) Experiments. [It was ascertained in the course of the proceedings, by both prosecution and defense, that the correct translation of "Fleckfieber" is typhus. A finding to this effect is contained

Figure 4.1 (Continued)

in the judgment. A similar initial inadequate translation occurred in the case of "typhus" and "paratyphus" which should be rendered as typhoid and paratyphoid.] From about December 1941 to about February 1945, experiments were conducted at the Buchenwald and Natzweiler concentration camps, for the benefit of the German Armed Forces, to investigate the effectiveness of spotted fever and other vaccines. At Buchenwald, numerous healthy inmates were deliberately infected with spotted fever virus in order to keep the virus alive; over 90 percent of the victims died as a result. Other healthy inmates were used to determine the effectiveness of different spotted fever vaccines and of various chemical substances. In the course of these experiments, 75 percent of the selected number of inmates were vaccinated with one of the vaccines or nourished with one of the chemical substances and, after a period of 3 to 4 weeks, were infected with spotted fever germs. The remaining 25 percent were infected without any previous protection in order to compare the effectiveness of the vaccines and the chemical substances. As a result, hundreds of the persons experimented upon died. Experiments with yellow fever, smallpox, typhus, paratyphus [It was definitely ascertained in the course of the proceedings, by both prosecution ad defense, that the correct translation of "Fleckfieber" is typhus. A finding to this effect is contained in the judgment. A similar initial inadequate translation occurred in the case of "typhus" and "paratyphus" which should be rendered as typhoid and paratyphoid] A and B, cholera, and diphtheria were also conducted. Similar experiments were conducted at Natzweiler concentration camp.

(K) Experiments with Poison. In or about December 1943, and in or about October 1944, experiments were conducted at the Buchenwald concentration camp to investigate the effect of various poisons upon human beings. The poisons were secretly administered to experimental subjects in their food. The victims died as a result of the poison or were killed immediately in order to permit autopsies. In or about September 1944. experimental subjects were shot with poison bullets and suffered torture and death.

(L) Incendiary Bomb Experiments. From about November 1943 to about January 1944, experiments were conducted at the Buchenwald concentration camp to test the effect of various pharmaceutical preparations on phosphorous burns. These burns were inflicted on experimental subjects with phosphorous matter taken from incendiary bombs, and caused severe pain, suffering, and serious bodily injury.

Taken from ***Trials of War Criminals before the Nuremberg Military Tribunals under Control Council Law No. 10.*** **Nuremberg, October 1946–April 1949. Washington, D.C.: U.S. G.P.O, 1949–1953**

Figure 4.1 (Continued)

purpose of completing a skeleton collection for anatomical research (Harvard Law School, 2018).

All 23 people who were tried faced counts of: (1) conspiracy to commit war crimes against humanity, (2) commission of war crimes and (3) commission of crimes against humanity. Ten of them were also subject to a fourth count of membership of a criminal organization (in this case the SS). Whilst none were found guilty on the first count, nine were found guilty on the remaining three, five others guilty of counts 2 and 3 and one other solely guilty of count 4.

In considering the moral and legal legitimacy of the Tribunal's actions in relation both to this specific trial and the trials more generally, one of the issues that needs to be addressed is whether the counts of war crimes and crimes against humanity that were central to the indictments in this and the other trials were legally justifiable. Certainly, the war crimes indictments had a clear legal basis because they were used in relation to violations of international law regarding conduct within occupied territories, including Articles 46 and 47 of the Hague Convention of 1907 which required respect for family honour and rights, human life, private property, religious conviction and practice and the avoidance of pillage. Indictments for crimes against humanity for domestic conduct were specified in the indictments to be justified on 'violations of international conventions, of internal penal laws, of the general principles of criminal law as derived from the criminal law of all civilized nations and were involved in and part of a systematic course of conduct.'

This was a more problematic assertion since as it was noted at the time (Wyzanski, 1946, 1) there were, in fact, no international conventions that dealt with crimes against humanity at the time, not all the conduct concerned was inconsistent with domestic law at the time (not surprisingly since in considerable part the law had been 'captured' to Nazi ends), and simply relying on the inconsistency of acts with the criminal law of other civilized nations smacked of overreach with respect to national sovereignty. The most one could argue is that it was a count that could be 'located' out of broader propositions of international law. Specifically, it was effectively advancing the notion that

> there exists, to cover the most atrocious conduct, a broad principle of universal international criminal law which is according to the law of most penal codes and public sentiment in most places, and for violations of which an offender may be tried by any new court that one or more of the world powers may create.
>
> (Wyzanski, 1946, 2)

However, this is not a proposition that would satisfy most lawyers as it

> would fly straight in the face of the most fundamental rules of criminal justice – that criminal laws shall not be ex post facto and that there shall be

> *nullum crimen et nulla poena sine lege* – no crime and no penalty without an antecedent law.
>
> (Wyzanski,1946, 2)

At the Doctors Trial, one could point to the fact that technically all 16 convictions that were secured would still have been secured without reliance on it. However, this is not to say it did not have a substantive impact since it may have influenced the degree of sentencing applied to the 15 defendants found guilty of it – 7 of whom received the death sentence and the remainder of which all received terms of 10 years or more even after some of them obtained commutation. However, it is at least clear from a moral point of view that all 15 should have been aware that their conduct fell short of the most basic level of humanity.

Equally the assertion by the defence that the experiments and medical atrocities more generally were conducted consistently with the norms applicable in other developed legal systems had little merit. One of the basic functions of any normal legal system is to protect the body from intentional interference and a logical corollary of this is a requirement to legally justify all forms of medical intervention. In English law, one can find case law recognizing that logic going back to at least medieval times. For example, in *The Surgeon's Case Nottingham Eyre* [AD 1329] Lincoln's Inn, Hale MS 1 (1), f58; Br. Mus. Egerton MS. 2811, f. 218, an action for trespass for loss of sight was defended by reference to the plaintiff having consented to the surgeon's attempt to cure his eye with herbs and other medications. Meanwhile the much later case of *Slater v Baker & Stapleton* 95 Eng. 860, 2 Wils. KB 359 (1767) provides an illustration of the general need for consent where treatment is carried out under contact. In the US, the need to justify medical intervention was recognized in a range of cases in the nineteenth century and early twentieth century. The most famous is *Schloendorff v Society of New York Hospital* 105 NE 92 (NY 1914) which articulated a general right to self-determination for adults of sound mind with respect to medical intervention. Equally significant in the immediate context was *Bonner v Moran* 126 F.2d 121 (D.C. Cir. 1941) which emphasized that the rights of patients should also be protected in the specific context of non-therapeutic intervention.

Such norms can be related to the traditions of the interests of patients being put first within medical codes which has existed since at least the birth of the Hippocratic Oath. It is also notable that the importance of respecting human beings in the specific context of performing medical experiments on them had already been formally recognized. In discourse, one can find it emphasized within various late nineteenth century and early twentieth century tracts on research (see further Rothman, 1998), amongst them the works of Bernard (1865) and Osler (1907) who suggested that non-therapeutic research was unacceptable altogether and Withington (1886) who noted the possible conflicts between the interests of medical science and those of patients which he suggested must be resolved in favour of the rights of the latter which should be protected in a patient Bill of Rights.

Furthermore, in terms of research specific law, a basis for respecting human beings existed within German tradition itself. Prior to Germany being formed, one of the states that went on to become part of it, Prussia, had in 1900 issued a directive to all hospitals and clinics excluding minors and incapable adults from being subject to medical interventions other than for diagnosis, healing and immunization (Vollmann, 1996). Furthermore, in 1931 a circular of the Reich Minister of the Interior created guidelines for new therapy and human experimentation. These stressed, amongst other things, that new therapeutic research could only be carried out without consent if urgently required and not capable of being postponed because of the need to save life or prevent severe damage to health and added that non-therapeutic research (human experimentation) was under no circumstances permissible without consent. Whether the circular constituted law or mere guidance was a point of contention at the trial and has remained so since (Grodin, 1992, 128). However, the mere fact that it was in force at the time of the experiments can be said to be central to justifying the convictions as more being based on more than a post-facto imposition of standards – especially when one considers that the ten principles that were laid out in the judgment as the appropriate measure for acceptable human experimentation (and subsequently became known as the Nuremberg Code, 1947) were themselves grounded on the content of the circular. In sum one could say that if anything the trial was both justifiable and useful and that the real problem was not the fact that it was brought but rather the fact that it was not accompanied by sufficient other efforts to root out and bring accountability for experimental and other medical abuses.

Taken on a global level, the trial and the Code have had a significant impact on the culture of medical practice generally and medical experimentation specifically over time. However, deficits in political will with respect to preventing, uncovering and bringing accountability for abuse have stunted there impact. The next two sections illustrate the problem when it comes to the US and Japan specifically. Between them they highlight the fact that much as the trial may have been justified it was still also an exercise in US governmental hypocrisy in as much as before, during and since the war the US government has allowed and in many cases sponsored extensive medical abuse and also decided to give immunity and in some cases jobs to leading figures responsible for Japan's World War II programme of research atrocity in exchange for their data. As I will now highlight, there were also weaknesses when it came to Germany's own approach to dealing with experimental atrocity and medical abuse more generally.

As an exercise of bringing wrongdoing German physicians to account, the Doctors Trial can be seen as doing little more than slicing off the very top of a very large iceberg. This is not a criticism of the trial itself – since as a war crimes trial it could only ever be expected to deal with the most heinous wrongdoers that had at this stage been identified. However, it is an important point to make since it highlights the need to view the trial as just one segment of an extensive patchwork of needed measures.

One of these other measures was domestic trials which though forthcoming to an extent were not routine and not always effective. The case study of Otmar

Freiherr von Verschuer, usefully illustrates the sort of accountability deficit that this could result in. Verschuer had provided a (pseudo)scientific foundation for Nazi racial theory and having taught Mengele went on to collaborate with him extensively, including via receiving a considerable number of body parts that he had taken from murdered Auschwitz inmates (Degen, 2014, 156). His activities were exposed after the war ended, but whilst he was prosecuted in 1946, the charge was minor, and though found guilty, he escaped with a fine (Hammerstein, 1989, 658). By 1951, he was back in a university position despite never having even apologized for his actions (Degen, 2014, 161). It has not escaped notice that his treatment was more akin to that of the Japanese experimenters shielded by the US (see Section 4.4) than to his fellow 'desk murderer,' Prof Dr Gerhard Rose who was sentenced to life at the Nuremberg Trials (later commuted to 15 years). Unwarranted positive input about his character from his former employer, the Kaiser Wilhelm Society, and the Confessing Church to which he belonged are perhaps the key to explaining the leniency of his treatment (Degen, 2014, 161).

The German system is also illustrative of the broader obstacles that can be met by those who are trying to hold a mirror up to a society generally and the medical profession more specifically to get it to confront, address and move on constructively from the full ugliness of its own past. The West German Medical Association (the forerunner of the German Medical Association) tried to employ the 'bad apple' theory for decades, suggesting from the late 1940s that 'Nazi medicine' was the responsibility of 350 rogue doctors and that most medical professionals had acted conscientiously and ethically in their duty to patients in the period of the Nazi regime even under the extreme pressures generated by it (Kolb et al., 2012, 722). Against the backdrop of a continued threat of revisionism of this kind, it is necessary to more fully elaborate the broader grounding that Nazi medicine had in wider medical and social movements. Many nation states were swept by a rising tide of eugenic thought toward the end of the nineteenth and beginning of the twentieth century (Bashford & Levine, 2010, 4). The notion that human beings could be ranked in terms of worth was widely prevalent in the period, and the idea that science could and should be used to give expression to this notion became increasingly popular not just amongst scientists but also policy and lawmakers and society as a whole. Radically exclusionary medical practices started to be given their head on a considerable scale (Kuhl, 2002).

One important example is sterilization of classes of human beings deemed socially unfit to reproduce. In the US, 17 states had created programmes for this by 1917, and by 1937, a further 20 states had joined them (Ferster, 1966, 591). In the same period, they were also set up in a range of other jurisdictions such as Sweden (beginning in 1906), Alberta and British Colombia in Canada (via The Sexual Sterilization Acts of 1928 and 1933, respectively) and Germany itself (via its Sterilization Law of 1933 which came into force in 1934). Germany's scientific community and society embraced the relatively benign notion of *rassenhygiene* in the late nineteenth century as part of an emphasis on societal betterment (Ploetz, 1895) but went on from the early twentieth to embrace and supercharge the international trend toward more disturbing eugenic ideals and practices. The authoritarian character of German institutions and familial relations (Reich, 2013) and the growing power of right-wing strands of thought were conducive to this as was the fall of German democracy and subsequent rise to power of the Nazi regime in 1933. Nazi ideology encouraged the notion that the Aryan race was superior to others. It also regarded certain classes of human being – not least gypsies, Jews, homosexuals, black people and the mentally disabled – as so inferior as to be subhuman. Whilst the principal responsibility for the victimization of such classes must be laid at those responsible for policy and practice within the Nazi regime, other parts of German society, including the medical profession, must bear shared responsibility for at least certain aspects of it as well. By the time Hitler came to power in 1933, 2,800 physicians had willingly joined up to the National Socialist Physicians League which had been set up by several physicians in 1929, specifically to co-ordinate Nazi medical policy (Gareth Jones & Whitaker, 2009, 74). Doctors also flocked to join the Nazi party more rapidly, and in larger numbers, than any other profession. By 1942, around half of them were members and around one-seventh of those who joined also became members of the SS (Proctor, 1992). Taken as a whole, the profession could be said to have helped to drive Nazi medicine as much as being driven by it, and the full horrors of how this moral stance had infected its approach not just to research but medicine more generally were there for all to see:

> doctors initiated and implemented an estimated 350000 coerced sterilizations, the killing of some 260000 people with mental illness or disabilities, and an estimated 25000 human experiments that led to the deaths of more than 2000 research subjects. Doctors selected individuals for slave labour or death at Auschwitz, and took part in the development and use of methods to kill people, such as by poison gas, fatal injections, starvation diets and electrocution. Tens of thousands of bodies of the executed were delivered to German medical institutions for training and research.
>
> (Kolb et al., 2012, 723)

Nazi medicine was clearly endemic to the medical profession as a whole with scientists of other kinds also extensively wrapped up with certain elements of the abuse, such as experiments conducted with chemical agents (Schmaltz, 2017, 253). Energetically and to some extent substantively, it was allowed to live on for many decades after the war. Insult was added to injury by the fact that right up until 1990 some German institutions continued to use the bodies of executed prisoners that had been delivered to them in the Nazi period (Kolb et al., 2012, 722) and by the fact that various medical professionals continued to draw on the data obtained from experimental atrocities committed during the period. Survivors of medical abuse and historical accounts critically exposing the breadth of wrongdoing within the medical profession were often marginalized (Kolb et al., 2012, 723), and when medical professionals put their head above the parapet by researching and disseminating wrongdoing, they did so at the risk of damage to their careers and even censure by the Association.

This mentality of suppression was not surprising. Post-war Germany made efforts to root out some of the baddest apples, but its medical establishment continued to be dominated by wrongdoers. A good case study of this was Hans-Joachim Sewering, a doctor who on his death in 2010 was praised by the Association for his use of great knowledge and skill to maintain the freedom and independence of the medical profession. Sewering was head of the Bavarian Medical Association from 1955 to 1991 and the German Medical Association from 1973 to 1978. In 1992, he was designated chairman of the World Medical Association but withdrew following accusations about his wartime role (Childs, 2010). Sewering had joined both the Nazi party and the SS prior to the war and following his graduation in 1941 served a year on the Eastern front. In 1942, he was assigned to Schonbrunn Sanatorium to work as a physician and was promoted to be its senior physician in 1944. It is estimated that whilst working there he signed orders to send around 900 disabled children to Eglfing-Haar camp. This was a centre for mass extermination that was euphemistically described as a 'healing centre.' Sewering died still insisting he had not known he was sending the children to be euthanized. However, this conflicts with testimony given by four nuns who worked their during the period (Roberts, 2008) and, in any event, seems unlikely given the simple fact euthanasia was an official element of Nazi policy.

4.4. Japan's warfare experimentation and use of chemical and biological agents

The use of chlorine gas by the Germans at Ypres on 22 April 1915 sparked Japans interest in developing a chemical weapons programme. Its Army Technology Review Board began to investigate the potential of developing an array of chemical weapons, means for their delivery and defensive measures that very year with Koizumi Chikahiko, a physician assigned to the School of Hygiene at the Army Medical College in Tokyo, emerging as the lead researcher and ultimately earning the moniker of 'father of chemical warfare in Japan' (Grunden, 2017). This facility was later complemented by two others that also carried out chemical warfare research – the Army Institute of Scientific Research which was established in April 1919 and the Navy Technical Research Institute which was established in 1923. Industrial-scale production of chemical agents had started by 1927. The 1920s also saw Japan taken an increasing interest in biological warfare. Japan was a signatory to the Geneva Convention, but its Diet had never ratified it specifically in order to be in a position to exploit the potential of such warfare. A special R&D facility for it, euphemistically named the Epidemic Prevention Research Laboratory, was set up at the Army Medical College. Personal accounts have supported the notion that some bacteriological experiments were conducted there on human beings (McCurry, 2011). When dead human beings were excavated from ground nearby in 1989 (Wang, 2014, 47; McCurry, 2011), the Japanese government denied that they were victims of such experiments, but this denial must be read in context of the fact that at this stage the government was still also denying that Japan had even operated a biological warfare related experimentation programme. In 2011, further excavations were started on the medical college site itself (McCurry, 2011). The facility was run by Dr Ishii Shiro (1892–1959). Ishii was a member of several of the secret societies that influenced the military and enjoyed a meteoric rise through its ranks in the 1920s aided by the patronage of the Japanese High Command.

Ishii transferred his base of operations to Manchuria (now northeast China) in 1932, following Japan's invasion (Harris, 1999, 2). In the village of Beiyinhe, South of Harbin, he set up a feasibility unit called the Togo Unit which had around 10 doctors and 100 other staff. It started to conduct bacteriological experiments on human beings by Autumn 1933. By May 1936, the Ministry for War had begun to put plans in place to enable work in Manchuria on both biological and chemical warfare to start to be greatly scaled up after a memo from the Army Chief requesting this had been specifically approved by Emperor Hirohito. The scaling up included inauguration of two units for biological weapons development – Unit 731 run by Ishii and Unit 100 – and one for chemical weapons – Unit 516. By 1939, Unit 731 was operating out of a specially designed complex in the Manchurian city of Pingfan (near Harbin) that consisted of around 3,000 personnel (Khabarovsk Trial, 1950, 8) and a sprawl of buildings that according to Harris (1999, 2) number 150 and including 2 secret prisons and 3 crematoria. Further units ended up being created for virtually every other area Japan conquered in the 1931–1945 period. Each specialized in one or a cluster of aspects of biological, chemical or general experimentation.

The human beings who were taken by the military and secret police to these Units to be experimented on were largely Han Chinese but also Koreans; Soviet prisoners of varying ethnic backgrounds; and, occasionally, Europeans, Americans and Australians. Along with allied prisoners of war, testers used captured communist partisans, ordinary criminals, political dissidents, those rounded up for suspicious activities, pregnant women, infants, elders and those who were mentally handicapped but physically fit (Harris, 1999, 2). When these were in short supply, the secret police would deliver poor and homeless people from the streets in cities throughout occupied China and Manchuria (Harris, 1999, 2).

As with the Nazi programme, the types of experiment that were conducted and the conditions under which they were conducted are indicative of the depths of depravity involved. General experiments included the likes of:

- Vivisection and removal of foetus from pregnant women.
- Limb freezing followed thawing (to study the effects of untreated gangrene and rotting) or amputation (which has also carried out separately to study blood loss and sometimes followed by grotesque reattachment of the limb to the wrong side of the body (Hudson, 2007).

- High temperature exposure (to discover the relationship between this, burns and death).
- Exposure to high voltage electric shock.
- Removal of essential organs for study.
- Hanging upside down to see how long it took to choke to death.
- Injection of animal materials such as horse urine into kidneys.
- Testing how long it took to die from such experiments as being starved and dehydrated, blood draining, being placed into hyperbaric chambers or being placed into centrifuges and spun.
- Exposure to lethal doses of x-ray radiation.
- Injection of sea water to determine if it could be a substitute for saline (Hudson, 2007).

Typically those who were experimentally subject to biological agents were infected directly and via explosions with diseases that had the potential to be weapons – such as anthrax, typhus, cholera, plague and glanders. However, some were infected with sexually transmitted diseases, such as syphilis and gonorrhoea, in order to try and better understand them to the end of reducing their prevalence amongst the Japanese forces.

Those who were taken to be experimented on at the Units were referred to by their captors as *maruta*. This description of them as logs or lumber may have helped to preserve secrecy but also reflected the fact that like the victims of the Holocaust, they were not regarded as human beings. Yudin (2014, 68) highlights an extract from the Khabarovsk Trial that can be seen as instructive given that human beings typically treat animals as of limited worth: When the defendant is asked what he did with the corpse of a Russian subject whom he had killed with potassium cyanide, he responds by saying that he dissected him at the cattle cemetery and under further questioning reveals that he subsequently buried the subject alongside already buried cattle carcasses. Whilst some victims were killed by the conditions they were kept in and some by the experimentation itself, many of those who survived both were then exterminated once their function as a subject was used up. Dissection of subjects was common. In the case of the biological experiments, subjects were killed and dissected at diverse points post-infection process in order to understand the infective process. Some Japanese veterans have also testified that for reasons of rigour and in some cases fun they killed subjects through rather than prior to dissection – dismembering them organ by organ *without anaesthesia* (BBC, 2005). A recently admitted example of this is eight American airmen who had variously been tested for the effects of ingestion of seawater, the effects of removal of parts of the liver and brain dissection (O'Flynn, 2015). Records of the Khabarovsk War Crimes Trial of 1949 show that, just between 1940 and 1945, 3,000 people were killed at Unit 731 alone (Khabarovsk Trial, 1950, 8). In this light, it is reasonable to estimate that 10,000 to 12,000 people may have been killed across the Units as a whole (Harris, 1999, 2).

Researchers at units such as Unit 731 also collaborated in the use of bacteriological and chemical agents in the arena of active warfare. Entire city and village populations were attacked with them typically via bombing (Kleinman et al., 2014, 1). The water supply in over a thousand wells in Manchuria was deliberately infected using cholera, typhoid and dysentery (Harris & Paxman, 1982, 78; Tsuneishi, 2014, 26–30). From 1940 onwards, there were several documented incidences of substances with plague attached being dropped in air raids on cities along the Yangzi river, with resultant outbreaks killing hundreds of people (Tsuneishi, 2014, 26–28).

From the beginning of 1942, pathogens such as cholera were in several cases dropped on battlefield zones, but this had mixed results. In the case of the Zhejiang campaign, it actually caused an estimated 10,000 Japanese casualties (Tsuneishi, 2014, 28). The Japanese were very successful in using bacteriological bombs to kill the easy targets of *maruta* that had been tied to stakes. However, they did not find the use of such bombs and other methods anything like as effective for killing defensively minded civilian populations, let alone highly mobile and well-organized opposition forces. The fact that the biological warfare programme was so readily persisted with may thus relate less to the potential for their effective use than with sadistic desires and the maintenance of professional empires. It perhaps also has to do with the interconnections between the programme and Japanese medical schools, hospitals and pharmaceutical companies. These not least included the fact that the programme provided plenty of opportunities to learn via general as well as biological and chemical agent experimentation, open up possibilities for learning surgical skills via vivisection and supplied a stream of brain cortex and other tissue that could be used for research and the development of profitable products (Wang, 2014, 45).

As the Japanese began to lose the war, some of their forces were captured by 'liberating' armies. When the Soviets captured those who they believed had played significant roles in committing atrocities through the Units, they made a specific point of charging them with war crimes. Twelve were successfully prosecuted at the Khabarovsk War Crime Trials of 1949, including the leader of Unit 100 Takahashi Takaatsu and some of his subordinates (Gomer et al., 1981; Yudin, 2014, 62–63). The Chinese authorities subsequently also mounted prosecutions of 45 suspected war criminals at the Shenyang and Taiyuan military tribunals. Nine were suspected of being tied with Unit 731 (five of them were convicted). The remainder were a mixture of members of others Units and doctors who had conducted vivisection on some of the local populace at local hospitals (Wang, 2014, 42).

The findings of the Khabarovsk Trial have largely stood up to subsequent analysis (Williams & Wallace, 1989; Harris, 2002; Nie, 2004). The US Government seems to have had its own dishonest political reasons for denouncing it as an exercise in communist propaganda. In 1946, the US had set up the International Military Tribunal for the Far East (also known as the Tokyo War Crimes Tribunal). The stated goal of this Tribunal was to try not just those who led the Empire of Japan to wage war, but also those who committed atrocities and crimes against humanity and those high up in the command structure who planned, ordered, authorized or failed to prevent them. However, most of those on US controlled territory who had significant roles in the Units avoided trial (for a discussion of exceptions, see Wang, 2014, 32–58) as did Emperor Hirohito in spite of his support of them and his role as the ringleader of the war. Indeed, he was allowed to continue being Emperor which buttressed the official and popular mischaracterization of him as just a figurehead (Nish, 2000, 86). The US had already secretly made the decision to secretly pay some of the experimenters for their research data and afford them immunity from prosecution in a cover up that it has never subsequently apologized for. Those benefitting from this kind of deal included Ishii, Kitano and other experimenters from Unit 731 (Guillemin, 2005, 89). Harris (2002, 264) illuminates on the reasoning behind this by quoting from the Dec. 12 1947 summary report of Dr Edwin Hill, Chief of Camp Detrick (headquarters of the US biological warfare programme):

> Evidence gathered in this investigation . . . has been obtained by Japanese scientists at the expenditure of many millions of dollars and years of work. Information has accrued with respect to the human susceptibility to those diseases as indicated by specific infectious doses of bacteria. Such information could not be obtained in our own laboratories because of scruples attached to human experimentation. These data were secured with a total outlay of ¥ 250,000 to date, a mere pittance by comparison with the actual costs of the studies.

In its gross utilitarian calculus, the US Government prioritized cheap, quick and easy data acquisition for its own biological weapons programme over the negative embarrassment which would ensue should their actions be found out. It seems that, on the one hand, the US Government seems to have been hoodwinked by the Japanese experimenters into thinking the data was more useful than it actually was (Bärnighausen, 2014, 82–83). On the other, none of the minutes, notes and other records of meetings that US authorities had points to them ever considering the ethical issues raised by prioritizing the potential usefulness of the data for their own biological weapons programme (Harris, 2002, 305). In a more respectful and less hypocritical world, the US Government would have exposed the wrongdoing to the cleansing effects of sunlight and made sure that ringleaders were punished at the Tribunal it set up or subsequently. It would also have supported efforts of victims and their families to get compensation, encouraged individual wrongdoers and Japan's Government and society to atone and intensively helped Japan to chart a course beyond cruelty and the related excesses of hierarchy and nationalism. Instead, as it was:

War criminals, free to resume their pre-war careers, dominated Japanese medicine and scientific research for a generation. These killers of innocent human beings rose to be Presidents of Universities, deans of Medical Schools, outstanding research professors whose scholarly research attracted world attention and praise in the scientific community.

> Many of the graduates of Unit 731 and the other Units controlled the Ministry of Health and its agencies such as the National Institute of Health (NIH). Former members of Unit 731 dominated the Board of The Japan Medical Society (the JMA) for many years. Other war criminals, such as Naito Ryoichi and Kitano Masaji, formed important pharmaceutical companies. The Green Cross Company, headed by Naito is perhaps the most important of these companies.
>
> (Harris, 1999, 4–5)

The Green Cross company was set up in 1950 as a private blood bank by Ryoichi Naito, a bacteriologist experimenter who had been instrumental in brokering the US immunity deal, and Dr Kitano Masaji, Ishii's right-hand man. In the mid-1980s, officials from the company, in collusion with members of the Ministry of Health, continued to supply patients with unsterilized blood after being warned that this would expose patients to the risk of contracting HIV (Harris, 1999, 5). Four other pharmaceutical companies were also implicated in the wrongdoing (Japan Times, 2000). This callous prioritization of company profit over patient safety resulted in thousands of people unnecessarily contracting HIV, some of them going on to die from AIDS as a result (Pollock, 1997). A 1996 Court mediated compromise forced the companies to pay 45 million yen (around 400,000 US dollars) to each victim as well as monthly payments of 150,000 yen (around 1,300 US dollars) to anyone who went on to develop full-blown AIDS, but it seems quite deficient that ringleaders were only charged with professional negligence and punished with short prison sentences rather than being charged with homicide and punished with long sentences (Japan Times, 2000). Another impact of the lack of individual and collective account for the atrocities was that the poisonous nature of Japanese racism that underpinned them was left largely unchallenged. Traditional Japanese notions of pureblood started to be mixed with US and German eugenic thought by the end of the nineteenth century (Otsubo, 2005). Japanese were discouraged from marrying outsiders, and by 1940s sterilizations began with a view to eliminating the reproductive potential of those deemed unsuitable to pass on their genes by dint of reasons that ranged from genetic disorder to leprosy. Such socially grounded sterilizations continued until the 1990s and have affected an estimated 16,500 people (Matsubara, 1998). In an echo of the warfare based experimentation situation (see below) victims have so far been denied government apology and compensation.

Japan's post-war failure to publicly highlight and account for its warfare experimentation is a 'triumph of inhumanity' (Nie, 2004, 32) and combined with its reckless post-war embrace of its callous experimenters has had a notably significant poisonous effect on its specific approach to research ethics. The culture around the use of human beings in such research entails a significant degree of disregard for their body related interests, expressed not least through insufficiency of commitment to principles of consent and proper control of risk (see further Figure 4.2). Harris has noted how this influence has continued beyond the first post-war generation of doctors (1999, 5).

Taking up this point further, Tsuchiya (2003) notes how contemporary understanding of research ethics in Japan is hardly regarded as a matter of 'human experimentation' ('Jintai-Jikken') which is a term used pejoratively in

In March 1951, Professor of psychiatry, Hiroshi Utena of Tokyo University published a paper titled "Studies on the Carbohydrate Metabolism in Brain Tissue of Schizophrenic Patients" [Utena & Ezoe (1951)]. In this study Utena let a surgeon take small pieces of brain tissue from 70 inmates of a psychiatric hospital. The subjects consisted of 42 schizophrenic patients, 21 "controls" including psychopaths, manic-depressives, compulsive neurotics, postencephalitic character abnormalities, epileptics, and general paralysis, and 7 "diagnostically questionable" patients including two "paranoid-hallucinatory" drug abusers. At least three subjects died after the lobotomy, and medical records testified that two of them probably died of excision. No subjects and their families were asked to give informed consent to excision. Several patients were forced to have general anesthesia against their will. Nevertheless, nobody criticized this experiment until 1971, when a resident Dr. Kiyoshi Ishikawa of Tokyo University Hospital accused Utena to the Japanese Society for Psychiatry and Neurology. Finally, JSPN's General Assembly resolved to condemn Utena in terms of possibly harming subjects and not attaining consent.

From November 1952-January 1956, Professor Juko Katsura of Niigata University and his disciples injected rickettsia tsutsugamushi to 118 inmates of a psychiatric hospital. The subjects suffered high fever with temperatures of around 40°c. Doctors also removed a piece of skin from 9 subjects. 8 subjects died after these experiments, one of them killed himself. Several employees of the hospital accused Katsura to the Human Rights Protection Committee of the Japan Federation of Bar Association, and JFBA set up a special committee and investigated the case. Katsura and his disciples insisted that injections were fever therapy, but it was revealed that there were no medical records, no fee was demanded for this "therapy," all injections were done by Katsura and his disciples, and Katsura gave a letter of gratitude to the hospital. In 1957 JFBA's special committee declared that it was surely an experimentation, not therapy, and severely denounced Katsura [Nichibenren (1968), pp.126-134]. This Katsura's experimentation

Figure 4.2 Examples of post-war human experimentation abuses in Japan.

was a part of a big research project on rickettsia funded by the U.S. Army [Takasugi (1973)].

In 1952, paediatricians of Nagoya City Medical School administered a special type of coliform bacillus to infant inmates of the Nagoya City Infant Nursery. The babies had severe diarrhea, and one of them died [Nichibenren (1968), pp.134-136]. In 1958, paediatric doctors of Kobe Medical School inserted a long plastic tube into infants from their mouth to anus, in order to take and analyze the contents of their intestines. That was a research procedure to know the best density of lactose in artificial milk, and doctors found that we must keep it less than 10%. But some baby subjects suffered from high fever, diarrhea, and vomiting. The parents and guardians of the babies were never informed of it, as the nursery completely shut them out [Shimizu (1979)].

Content adapted from Tsuchiya, 2003.

Figure 4.2 (Continued)

Japan to refer to the worst sorts of atrocities performed by evil doctors. He argues that whilst many Japanese bioethics texts refer to the likes of the Nuremberg Code and the Declaration of Helsinki, the 'heavy, important and essential' meanings that such codes have for Japan is missed as a consequence of relative lack of reflection on national past misdeeds. Some local authorities and citizen activists have made efforts to air the truth of the Japanese atrocities, such as by showing a 'Unit 731 Exhibition' in 61 locations in 1994–1995 which included testimony from some of the more 'repentant' members of that Unit (Gold, 1996). However, generally speaking post-war abuses (such as those noted in Figure 4.2) have tended to looked at as a case-by-case issue and reported in this fashion in the media rather than integrated with a more general discussion of research ethics and specifically related back to the systematic problems of Japan's past and to some extent continuing research culture. As a result, they have not exerted that much influence on research governance and practice.

Warfare experimental atrocities have exerted even less influence on research governance practice. The desire to avoid confronting a shameful past may be one factor here, but there are also undoubtedly others. Members of medical academies who come to know the details of these atrocities may often have been fearful that speaking up about them will jeopardize their careers, particularly given that some of those responsible for the atrocities have been occupying positions of power.

No doubt partly to avoid new lines of political, moral and even potentially legal accountability, Japanese Governments have also shamefully stood in the way of these atrocities being properly addressed. For example, as late as 1996, the Japanese Government was continuing to falsely claim that it had not found any official documents confirming evidence that germ warfare was employed in

China, despite having thousands of such documents in its archival holdings (Harris, 1999, 4). When a suit, lodged against it by Chinese victims and their families, was heard in the Tokyo District Court in 2002, its role was exposed by the judges own fact-finding mission (Jones & Simkin, 2002). However, whilst it was then effectively forced to accept the facts of the situation, it refused to compensate the victims and their families and the Court refused to compel it to do so, reasoning that Chinese people have lost their rights to judicially claim war compensation from Japan, Japanese people or its companies under the 1972 Japan-China Joint Communique (Schoppa, 2011, 301).

This seems to be somewhat disingenuous reasoning given that the Communique did not specifically refer to individual rights to claim and also given that Japan, at that time, had denied carrying out the wrongdoing in question. Nonetheless, the same reasoning defeated appeals to Japan's High Court in 2005 and its Supreme Court in 2007 (Schoppa, 2011, 301). As well as being unwilling to take proper collective responsibility for the atrocities, the Japanese Government has thwarted efforts to bring individual perpetrators to account by shielding their identity. For example, the names of members of Unit 731 were not disclosed by it until 2018 (McCurry, 2018) – 72 years after the end of hostilities and hence when nearly all of them can already be expected to be dead. For the most part, those who have already been publicly known have been unwilling to take their own share of responsibility. Some continue to defend what happened, and many simply do not wish to recall or speak about it. Guo (2014, 118) observes that only a handful have ever expressed guilt. However, taken together the accounts of these individuals speak variously to: shock and shame at the horror of what they have allowed themselves to become a part of and that the Japanese could do such experiments without guilt; deep sadness for the hurt that they have caused and the desire to be punished for it; the sense that their activities did not even serve any useful purpose since the experiments could have been conducted on animals instead and vivisection did not provide any knowledge of the body that could not be gleaned through textbooks; *and* the sense that, because of this, those activities must have been otherwise motivated, such as by curiosity and the desire to engage in human slaughter (for a collation of the accounts, see Guo, 2014, 111–118).

In sum, it may be concluded that the broad scope and scale of experimental atrocity under the Japanese programme is now relatively clear and its lack of respect for human worth is no longer feasible to deny. Nonetheless, the story of the programme has been partly one of how accountability for atrocity has and can be avoided and what a deleterious impact that can have not just on the victims but on the country whose Government refuses to take responsibility properly and whose medical establishment continues to provide a haven for perpetrators. It is also a story with elements that are yet to be publicly known or told, and in the coming years it would not be surprising to see more bodies being dug up both metaphorically and perhaps also literally.

4.5. The US World War II and cold war experiments

The US maintained a chemical and biological warfare programme over the course of World War II and conducted various human experiments as part of it. These included, for example, the Malaria Research Project which entailed doctors from the University of Chicago exposing 441 inmates to bites from malaria-infected mosquitos at Stateville Penitentiary in Joliet, Illinois. After the war had ended, it continued the programme. It hypocritically drew on knowledge from the Japanese programme whilst sustaining the myth of American innocence (MacDonald, 2014, 170) and even actively peddling the notion that it was the world's policemen not only for warfare atrocities generally but experimental ones in particular (such as via its leading role in the Nuremberg Trials and specifically the so-called Doctors Trial).

By the 1950s, it developed a $90 million USD biological agent mass production facility at Pine Bluff in Arkansas and had upgraded Camp Detrick and developed various specialized laboratories for the test of a considerable number of biological agents (Williams & Wallace, 1989). However, perhaps even more significant than this was the fact that it developed a radiological programme in the latter half of the war which it then continued through most of the cold war period. Through that programme, the US engaged in thousands of radiation experiments on human beings principally via the methods of injecting radioactive material, total body irradiation and the use of isotopes. Many of these experiments were abusive, and it can be said that as a whole the programme not only reflected a heavily prevalent lack of respect for human beings within medical research culture as a whole but also profoundly reinforced and strengthened it.

Experiments with injecting radioactive material started with plutonium being put into rats in 1942 by Joseph Hamilton at the medical school at the University of California with support from the Office of Scientific Research and Development. By 1943, the US had started to build up plutonium supplies on a larger scale pursuant to efforts at the Manhattan Project to develop an atomic bomb. When one of the project's workers was accidentally exposed to the entire plutonium supply of 10 milligrams, the impulse to conduct further injection experiments with radioactive materials gathered pace. Between 1945 and 1947, 18 patients from across several hospital sites were injected with plutonium. Other radioactive substances injected into human beings in small scale trials included polonium, uranium, radio-yttrium, radio-strontium and radio-cerium. As a general rule, the patients used in these trials derived no benefit and were being administered quantities of radiation significant enough to generate health damage, particularly in the long term. In spite of this, they usually did not know what was being done to them let alone freely agree to it. There are also hints in the evidence presented in the ACHRE report that the experiments reflected status based thought, with many chosen to be subjects on the basis of the reasoning that serious illness made them less valuable (ACHRE, chapter 5, parts 1 and 2).

When oversight for such experiments passed to the civilian Atomic Energy Commission (AEC) on 1 January 1947, it made the decision to keep nearly all the information secret. This was 'evidently for reasons not directly related to national security, but because of public relations and legal liability concerns' (ACHRE, chapter 5, part 3) and to maintain supply of research subjects (MacDonald,

2014, 171) which in turn suggests there was awareness that the experiments were inconsistent with ethical norms. Indeed, when interviewed about the matter in 1974, Shields Warren, who became director of the AEC's Division of Biology and Medicine in late 1947, said that it had been decided

> that the rules [should be] properly drawn up by the. . . . Human Applications Isotope Committee . . . so that use without full safeguards could not occur, and that . . . nothing of the sort could happen in the future.
>
> (Miazga et al., 1974, 10)

These safeguards included consent meaning that the AEC was putting itself in the hypocritical position of demanding a transparent approach whilst simultaneously seeking to hide past abuse from public view.

By the end of 1947, the inherent limitations of this philosophy of attempting to reform without exposure to the sunlight of public view were already evident; Warren put an end to isotope experiments on human beings at the University of Rochester but was persuaded to allow plutonium experiments on human beings at the University of California to continue. In his evidence to ACHRE in 1974, Warren attempted to distinguish the two by suggesting the latter involved consent, albeit not informed because the effects of plutonium were [relatively] unknown (Whittemore, 1995, 3). However, this is a patently distorting analysis at best. What the subjects were told, according to Warren's testimony, was that

> they would receive an injection of a new substance that was too new to say what it might do but that it had some properties like other substances that had been used to control growth processes in patients, or something of that general sort.

However, analogizing plutonium to substances that 'control growth processes in patients' was misleading rather than informing (ACHRE, chapter 5, part 4).

Much of the activity alluded to was problematic not only for the way it treated subjects at the time of experimentation but also after; many subjects were put to inconvenience and detriment in the course of follow-up tests and interventions. Some were subject to the indignity of exhumation. Virtually all who had been misled continued to be dishonoured by continued failures on the part of the researchers and related organizations to fulfil their duty of veracity and make good on their wrongdoing.

After the last of the injections at the University of California, abuse was perpetuated through the Boston Project which conducted human uranium-injection experiments from 1953 to 1957 at Massachusetts General Hospital (MGH) as part of a co-operative project between the hospital and the Health Physics Division of Oak Ridge National Laboratory. Eleven patients with terminal conditions were injected with uranium. The investigators found that the maximum permissible levels for uranium in water and air might be unsafe, but over the years, the US Government has avoided following its own recommendation for more conservative occupational standards. In addition, hopes that uranium could be used as a radioisotope that would localize in (hitherto untreatable) glioblastomas and destroy them when activated by a beam of neutrons have remained unfulfilled. Subjects were being given very large doses of uranium which data had proved to be very harmful, and on this basis the project was discontinued in 1957 (ACHRE, chapter 5, part 4). In 1995 the Advisory Committee was told by the lead investigator that consent was obtained from patients or their families and that they scrupulously gave patients all the information they had (ACHRE, chapter 5, part 4). This claim seems somewhat incredulous in the face of no documentary evidence having been identified to support it (ACHRE, chapter 5, part 4).

The ACHRE report has concluded that these injection experiments as a whole (with the possible exclusion of those involving polonium) were unethical (ACHRE, chapter 5, conclusion). Subjects and their families were treated in an egregiously disrespectful fashion heightened by the fact that they were ill and institutionalized in hospital, making them more vulnerable to exploitation.

> As patients, it would have been reasonable for them to assume that their physicians were acting in their best interests, even if they were being given 'experimental' interventions. Instead, the physicians violated their fiduciary responsibilities by giving the patients substances from which there was no expectation they would benefit and whose effects were uncertain.
>
> (ACHRE, chapter 5, conclusion)

The ACHRE Report also criticized the idea that it was legitimate to allocate the risk of harm by giving higher doses to 'terminal' patients, noting that whilst a person who is dying may have fewer interests in the future than a person who is not that this does not mean they are owed less respect and can be used like an object/mere means to the ends of others (chapter 5, conclusion). In other words, these patients should not be treated as exceptional but have their interests protected via the robust application and enforcement of normal rules.

Another category of problematic experiments involved total body irradiation (TBI). Also known as whole body irradiation, TBI involves the use of external radiation sources that produce penetrating rays of energy to deliver a relatively uniform amount of radiation to the entire body. From the late 1930s, it was understood that, whilst it could be administered at low dosages to treat lymphomatoid tumours (such as leukaemia and lymphoma), it had to be administered at life-threatening dosages to even provide temporary relief of carcinomas (for example, of the breast, lung, colon and other organs). Its subsequent use for the latter might have been presented as therapeutic but was not and was actually designed to fulfil hidden agendas.

From the early to mid-1940s, research teams at several sites conducted experiments on such patients to yield data on the impact of radiation exposure for the Manhattan Project. One such team, at the University of California Hospital, also used arthritic patients who very clearly could not benefit. Another, at the Chicago

Tumor Clinic, included three healthy volunteers as controls (Nickson, 1951). The AEC colluded with this abuse rather than confronted it. In 1948 it produced a report that conveyed the false impression that subjects had gained a good, if not optimal, therapeutic treatment through the experiments, but it cannot have been particularly convinced of this given that to avoid litigation it edited out information which people would need if they were to be able to identify themselves as having been subjects of the experiments (ACHRE, chapter 8, part 3).

Via five programmes of experimentation that were conducted on hospital-based cancer patients from the 1950s until the 1980s, the atmosphere of deceit was continued. The patients were not told that these programmes, sponsored by the Department of Defense, were designed to examine the effects of various levels of exposure for a variety of military purposes (ACHRE, chapter 8, part 5). Furthermore, four of them drew on patients with radiosensitive cancers. In some cases, subjects were drawn almost exclusively from public hospitals and were typically very ill, powerless and often African-American (ACHRE, chapter 8, conclusion).

Experiments using radioisotopes have also been the source of ethical problems. The medical potential of radioisotopes was recognized before World War II. New knowledge of the atom gained in the early to mid-1940s provided a springboard for the exploration of this potential after the war. By mid-1946, the Manhattan Project had set up an oversight and distribution system, and by October 1946, distribution for experimental use was underway. When the AEC came into being at the beginning of 1947, its Subcommittee on Human Applications of the Advisory Committee on Isotopes Distribution Policy took over the baton from the Manhattan Project. The subcommittee regulated types of uses allowed according to their hazard and the extent of our knowledge of the risks and requirements. It also required and provided training of those who would use radioisotopes and required the establishment of local radioisotope committees, which not only reviewed proposals but suggested changes at the local level in experimental design to reduce risk.

In practice, controls on risk were not as strong as they should be. This was particularly the case with vulnerable groups. The seriously ill were often treated as being more expendable by being given larger than normal doses. Minutes of the Subcommittee meeting of 13 March 1949 point towards the exclusion of non-therapeutic use of pregnant women, but this may well have subsequently occurred (ACHRE, chapter 6, part 3). The use of children was permitted without any emphasis being placed on the need to protect their best interests, in spite of the risks of even low radiation dosages. Furthermore, the Subcommittee's policy of discouraging and restricting participation of children was only applied to those deemed normal. This left the door ajar for abusive research to be carried out on those not deemed to be normal. Indeed, the AEC, along with the National Institutes of Health and the Quaker Oats Company funded one infamous abusive research project conducted by MIT. This entailed non-therapeutic administration of radioactive iron and calcium to the breakfast cereals to a number of students at the Massachusetts based Walter E. Fernald School in the late 1940s and again in the early 1950s. The School was officially for the 'mentally retarded' but was actually a state 'dumping ground' for many types of children (Task Force, 1994). Another infamous non-therapeutic study was funded by the Division of Radiologic Health of the US Public Health Service to test a proposed countermeasure to nuclear fallout by determining the amount of non-radioactive iodine that would effectively block the uptake of radioactive iodine that would be released in a nuclear explosion. Conducted in 1961 with researchers from Harvard Medical School, Massachusetts General Hospital and Boston University School of Medicine, this involved the administration of small amounts of radioactive iodine to 70 children at the Wrentham State School, another Massachusetts facility for what were then termed 'mentally retarded' children (The Working Group, 1994). Such studies were the worst end of the spectrum of abusive isotope experiments with children, but it is important to note that certain problems appear to have been more generic with respect to the isotope experiments, such as failure to gain parental consent (ACHRE, chapter 7, part 2) though doing so would not, in any case, have justified performing interventions on children that were clearly not compatible with their best interests.

With respect to the use of isotopes generally, there is no evidence that the AEC Isotopes Distribution Division applied the standards for documented consent AEC General Manager Carroll Wilson had articulated in 1947. A limited consent requirement was developed for the administration of larger-than-usual doses to very sick patients. However, only in 1956 did a consent requirement for normal volunteers appear in the AEC guidelines and this requirement was not sufficient to require the kind of detailed disclosure that is legally normative in the United States today (ACHRE, chapter 6, part 3). The same type of approach was continued in 1965 guidance (ACHRE, chapter 6, part 3). Indeed, it is likely that a proper approach to disclosure only became entrenched by the end of the decade as a result of local committees responding to a combination of public demand for a better approach and the development of federal requirements (ACHRE, chapter 6, conclusion).

In sum, the histories of experimental use of TBI, radiation injections and, to some extent, radioisotopes are closely interwoven. The histories of all three are littered with cases of subjects being used without respect for their rights to make an informed decision about whether to participate or not. The histories of the first two are littered with cases where levels of harm and risk are too high taken in isolation or when measured against the limited or absent potential for them to bring benefit to subjects of them.

The experiments often also discriminated against a number of categories of human being as a function of what appears to be a mix of a personistic ranking of human beings and sheer opportunism. The sick and children with disabilities were often specially selected for use as guinea pigs for harmful and risky experiments because they were perceived as more dispensable. Children and pregnant women were selected not just for their potential use value but also their exploitability. Servicemen were often selected for their exploitability. A mix of not just status based arguments and exploitability but also plentiful supply and low cost appears to have been at work with prisoners (Hornblum, 1997). Statistics relating to class and race were not routinely recorded at the time. However, as already noted, some of the experiments were conducted at public hospitals and other institutions such as prisons where the poor and non-white populations were highly over-represented. And with regard to race specifically, discrimination in the US was an even more endemic fact of life in the period concerned than it is today. Medicine reflected and advanced this problem by its cultivation and embrace of racial and eugenic theory and related enthusiasm for race based forms of extreme medical abuse, such as: routine non-consensual sterilization of Native American young women in Indian Health Service Hospitals (Ralstin-Lewis, 2005); the Tuskegee experiments which involved studying the effects of syphilis on around 400 infected African-Americans who were deceived into thinking they were being treated – this was prefaced by an earlier experiment in Guatemala from 1946 to 1948 where the US Government ran a project with the support of the Guatemalan president and some health ministries to infect various Guatemalans with syphilis; and the two sets of radiation experiments conducted largely on African-Americans at University of Cincinnati College of Medicine that have been mentioned earlier in this chapter.

As some of the evidence submitted to ACHRE highlights (chapter 7, part 2), a 'war on disease' was a key pretext for a lot of the experimental excesses. Individuals were 'drafted' to sacrifice their medical interests for this war in parallel first with the literal war of 1939–1945 and then the cold war. This is reflected in the fact that parallel failures to protect children were occurring in other areas of research. One infamous example cited by the ACHRE report (chapter 7, part 2) is the hepatitis studies conducted by Dr Saul Krugman of New York University in the 1950s and 1960s at the Willowbrook State School, an institution for the 'severely mentally retarded.' Krugman and his staff systematically infected newly arrived children with strains of the virus in order to study the natural history, effects and progression of the disease. Such experiments patently abused the subjects making it essentially irrelevant that parental permission was obtained. In any event that permission appears to have been granted without proper disclosure. There may also have been inducement – the allegation being that entrance to this overcrowded institution could be secured quicker through participation (Beecher, 1970, 122–127).

4.6. The UK cold war experimental programme

Like there US counterparts, the UK Government continued to run a secret weapons programme after the end of World War II and conduct a range of associated experiments. Among the experiments were over 750 'mock' biological and chemical warfare attacks launched from aircraft, ships and road vehicles (Schmidt, 2015). When these experiments have been exposed through government documents or otherwise the general official position has been to deny that any harm has been caused. However, this position has become increasingly untenable in many cases.

Some of the experiments entailed dispersion of zinc cadmium sulphide, which is now known to be potentially carcinogenic. Some others entailed the spread of harmful radioactive dust. For example, in 2002, the BBC discovered that the large clouds of gas that had been released on four separate occasions from the top of an incinerator stack at Harwell Nuclear research facility in Oxfordshire in July 1961 were in fact xenon (Thompson, 2002), the harmful radioactive properties of which are sufficient to generate a small risk of cancer and/or genetic damage to those inhaling it. In the experiments, it swept rapidly over large parts of Southern England. The relatives of Charles Wallington, a meteorologist involved in the Harwell experiments who later died of a rare blood disorder, believe his death may be linked to the work he conducted.

There were also many field experiments that involved dispersing, what were contended to be harmless biological agents in ways that would simulate harmful ones. For example, *Bacillus globigii* (now commonly known as BG) was dispersed on numerous occasions over large areas (typically more than 100 miles) covering significant numbers of people (up to half a million in some cases) in an effort to simulate how anthrax might spread if it were released. In 1999, in the wake of criticism of its use, the MOD commissioned Professor Brian Spratt Fellow of the Royal Society (FRS) to produce a report on its effects. On review of the primary literature, he suggested that no examples of significant disease were clearly attributable to it and noted that, whilst BG was now considered a variant of *B.subtilis*, which was occasionally encountered in hospital patients, its association with disease was still not clear. (Spratt, 1999). This analysis not only ignored the precautionary principle but ignored what was clear about *B.subtilis* from even standard microbiology texts of the time – namely that it was part of a group of bacteria that cause infections, particularly in newborns, those recovering from invasive procedures and those with immune system compromising conditions. BG specifically has now widely recognized to be a pathogen for some time (see, for example, Miller et al., 2004) and understood to be capable of causing such problems as fever, food poisoning, peritonitis and septicemia. A Porton Down Risk Assessment (Dstl/DET/G182) declassified in 2006, points to DERA having at some point decided that it had sufficient potential to cause harm to warrant precautions being taken to minimize exposure to it.

E.coli MRE162 was another clearly harmful agent that was dispersed on a regular basis over a long period, with Harris and Paxman (1982, 157–158) suggesting that it was used in 43 field trials between 1963 and 1971. The potential impact of release of agents in densely populated areas was also investigated, with a number of trial utilizing the London Underground. For example, on one occasion in July 1963, spores of bacteria, housed in the sort of compact cases you would find in most women's handbags of the day, were thrown out of tube train windows. Scientists measured that by tea time they had travelled up to 19 km across the capital's tube system (Thompson, 2002).

The Ministry of Defence also continued to use Porton Down as a site for testing chemical weapons on human beings. British servicemen were tested in secret until the 1960s and then more openly until 1989. The worst of these tests involved subjecting servicemen to sarin, a highly toxic nerve agent developed by

Germany during the course of World War II. One of the servicemen, Mr Ronald Maddison, died in 1953 just 30 minutes after having 200 mg of liquid sarin dripped onto a patch of uniform on his arm just 9 days after another serviceman James Kelly had suffered near fatal poisoning from a similar incident. Although none of the servicemen was literally forced to participate, their involvement was routinely procured on an uninformed basis. Indeed, some appeared to have been positively misled – for example, Mr Maddison's family claimed that he was under the impression that he was taking part in a test to research the common cold whilst another participant, Mr Foulkes, stated that whilst he was told that he was participating in a sarin trial he was misled as regards the prospective health implications of doing so:

> I specifically asked them what the long term implications of taking part in the tests were because I was not happy about it. Of course, if they had mentioned what happened to Ronald Maddison I would not have taken part.
>
> (Syal, 2002)

Harm and deceit were compounded by the obstacles that were placed in the way of servicemen and their families finding out the full extent of what had happened and obtaining redress for it. In the year of Mr Maddison's death, the then Wiltshire Coroner, Mr Harold Dale, conducted an inquiry into his death. However, we effectively colluded with the Ministry of Defence by holding this in secret and reached the shocking verdict of death by misadventure, in part aided by the fact that he 'was not provided with a full account of the circumstances' of the prior experiment, involving the near fatal prior poisoning of Mr James Kelly (per Lord Woolf, Chief Justice in *Re Maddison, Deceased* [2002] EWHC 2567 Admin at para. 15).

To their credit, the Wiltshire Police both started a criminal investigation into the abuses in 2000 and went on to conclude that three of the scientists may have breached section 24 of the Offences of the Person Act 1861 which makes it an offence to 'unlawfully and maliciously administer to or cause to be administered to or taken by any other person any poison or other destructive or noxious thing, with intent to injure, aggrieve, or annoy such person.' Mr Dale's decision was then reviewed at the behest of a new Wiltshire Coroner in *Re Maddison, Deceased* [2002] EWHC 2567 Admin. Lord Woolf, Chief Justice made the general point that 'it is important that there should whenever practicable be a public, full, fair and fearless investigation as to the causes of death' (para. 10). He concluded that there was material 'which was not properly and fully investigated at the previous inquiry' (para. 12) and that there were three exceptional circumstances which together could justify the quashing of the original verdict, namely that: The death had occurred at the hands of the state in the context of an experiment that risked life (para. 13); in spite of the passage of time making it difficult to hold an investigation, substantial records still existed (para. 14); and finally, the previous inquiry had been held in secret meaning the family of Mr Maddison had never had the opportunity of hearing a public investigation into the causes of his death (para. 15).

A fresh hearing saw a jury return a unanimous verdict of unlawful killing on the 15th of November 2004. Following this, the Government changed tack and settled claims made by servicemen and their families in relation to LSD experiments in 2006 (BBC, 2006) and then in relation to sarin in 2008. However, both settlements can be criticised as belated and the latter was accompanied by a half-baked apology that added insult to injury by engaging in the moral relativism of emphasizing the cold war context in which the experiments had taken place (BBC, 2008). Furthermore, no prosecution has been mounted in relation to the sarin experiments because the Crown Prosecution Service has adopted the seemingly incredulous position that there is insufficient evidence against scientists who were involved in the wrongdoing (CPS, 2011). In spring 2018, the MOD added further insult to injury by continuing to downplay the significance of Ronald Maddison's death. The MOD called it 'a tragic and regrettable incident' and maintained its stance of prioritizing supposed expedience over respect for human life by opportunistically defending the 'volunteer' programme as having facilitated the development of protective clothing and medical countermeasures deployed in incidents such as the one it was then being confronted with in Salisbury stemming from what has been officially pronounced to be the use of Novichok on the home of Sergei Skripal in Salisbury (Wright, 2018).

4.7. Warfare related medical use in post-war autocracies

The fact that warfare medical use was pursued after the end of World War II in some autocratic jurisdictions is less surprising and exemplative than the fact that it was also pursued in some democratic jurisdictions, such as the US and the UK. Nonetheless, it is worthwhile reflecting on. One of the significant features of the post-World War II period was how quickly matters descended into a cold war with the Soviet Union and its allies on one side and NATO countries and their allies on the other. One feature of this cold war was a race to develop military might including nuclear, chemical and biological weapons that had enough capacity to destroy human and at most other forms of life across the world many times over. With the collapse of the Soviet Union in 1992, there was an effort via the Nunn-Lugar programme to ensure that new nations inheriting staggering quantities of nuclear, chemical and biological weapons and the infrastructure that supported them were provided support to safeguard and destroy them. The destruction aspect of this work was, however, never total. Furthermore, the principal new nation, the Russian Federation, has retained significant stockpiles and an active research programme. In spite of signing the 1993 Chemical Weapons Convention and ratifying it in 1997, it draws on its chemical weapons to carry out targeted assassinations. Some of these have been carried out on foreign soil. For example, in 2006, radioactive polonium-210 was used to fatally poison Alexander Litvinenko in London, and in 2018 the exceptionally powerful nerve agent Novichok was used in Salisbury to try and kill the turned former Russian spy Sergei Skripal. This latter mission was allegedly carried out by two Russian military officers (one of whom is a doctor). It not only caused serious harm to Skripal and his daughter, but also collateral damage to others, including Dawn Sturgess who ultimately died from exposure after her partner had picked up and given to her what looked like a perfume bottle that had Novichok in (Greenfield, 2018).

North Korea is another example of an autocratic regime that has been active in this field. Having been formed in 1948, it has been active in the atomic field since the 1950s. Testimony from a range of defectors also suggest it has engaged in gruesome forms of warfare related experimentation. These range from duping people into eating poisoned cabbage to studying the effects of starvation, testing lethal gas on human being in gas chambers (BBC, 2004; Ryall, 2017). There is a concern that, if North Korea starts or enters into a war with the South and/or other states, it may use chemical and biological weapons as part of its armoury of tactics. Like the Russian Federation, it appears both willing and capable of effectively deploying chemical agents for the purpose of targeted assassination. The killing of the current President's brother Kim Jong-nam at Kuala Lumpur airport in early 2017 appears to be a case in point. The Malaysian police have confirmed that he died through being poisoned with the highly toxic liquid nerve agent VX (Holmes & Phillips, 2017). The police reported that he was smeared on the face with a chemical compound by two women, and that later on (as confirmed in CCTV footage) another woman approached him from behind and covered his face with a textile cloth of some sort. It seems likely that this cloth was used to place a further chemical compound on him and that this reacted with the first compound to produce VX.

Some other autocracies are known to have started and drawn on programmes in the period since World War II ended. Use of chemical weapons was one of the means by which Saddam Hussain's Iraqi regime prosecuted war against Iran and also one of the means he used to attack parts of the Kurdish population both in Iran and within Iraq's own borders (Mlodoch, 2017). Assad's Syrian regime is also alleged to have used chemicals such as chlorine and sarin repeatedly on parts of its own population (Trapp, 2017), and Daesh appears to have used mustard and potentially other gases as part of its efforts to take over parts of Iraq and Syria (Trapp, 2017, 372).

4.8. Conclusion

In World War II alone, 50 million people were killed. It is the destructive potential of war in general that should most concern us. Nonetheless, abusive warfare related medical use of the human being is significant not least because it provides an important and especially disturbing element of that potential. One of the things that is most disturbing about it is the incongruity of using medicine to destructive ends. Clearly, such use does not fit with how medicine is thought about in an everyday sense or with medical ideals found in medical codes such as the Hippocratic Oath. Nonetheless, I would argue that such use has a very close nexus with medicine indeed.

Part of this is down to the mundane fact that both warfare related chemical, biological and radiological programmes and warfare related experimentation in general rely heavily on professionals with medical backgrounds to effect their development and continuation. The other part is down to wider issues with the way in which medicine is being thought about and practiced. As noted in Chapter 1, the mechanistic nature of medical orthodoxy is such that it lacks a proper understanding and respect for purpose and potential. In promoting a sense of hopelessness amongst human beings about what they are here for and capable of, it ripens conditions for their objectification, dehumanization and general exploitation. Because of this and because it is not immune from more malign ideological forces that exist in society more generally, medicine has provided a climate that has tended to be somewhat conducive to the development of abusive practices including abusive forms of medical use of the human body.

One aspect of this that is particularly worth dwelling on is how it has proved to be a place for cultivation as much as rejection of the ranking and differential treatment of different classes of the human species, including via the development and propagation of eugenic theory and practice which as we have seen was used in perhaps its most devastating forms under the racist totalitarian Japanese and German regimes of the 1930s and 1940s. Williams and Wallace (1989, 37–38) highlight a motivational speech given at the first meeting of Unit 731 (possibly by Ishii himself). The speaker noted the God-given mission of doctors to prevent disease and provide the most expeditious treatment and acknowledged that what the Japanese programme was doing might cause some misgivings because it entailed the opposite. However, he assured them that their research would provide a 'double medical thrill' consisting of enabling them to not only to satisfy their interest in probing an unknown world but also to build a powerful military weapon for use against the enemy. Nie (2014, 125) places the interest in doing so into the context of nationalism having been closely entwined with Japan's remarkable pursuit of modernization and modernity, including not least via science and technology. The notion of moral discipline central to Shinto religion had already been heavily perverted to a predominant emphasis upon loyalty and filial piety to the nation, the state and the Emperor, suiting the ends of an increasingly expansionist militantly nationalistic and totalitarian regime. As Nie concludes, '(t)oo often, human beings, physicians included, have subordinated the human race to the nation forgetting the higher claims of human brotherhood' (2014, 135). Medicine has been used in a parochial way against the universal aims of treating and preventing disease that enable it to be a force for goodness and peace within and between nations.

The phenomenon of war based exceptionalism has also arisen in ways that go beyond the parochialism of nationalism. Sass, for example, expresses the sentiments of many defenders of such exceptionalism when he suggests that extreme situations require, extreme, sometimes very unpleasant and very extraordinary means (2005, 164). Contrary to that assertion and in line with the thinking presented from the outset of this book, I would agree with Döring that basic ethical principles should always be respected (2014, 142). Furthermore, even if there were a case for overriding them in certain circumstances, it would be hard to conclude that these circumstances would ever include conducting atrocious experiments and inventing, manufacturing, stockpiling and using chemical, biological and radiological weapons. It might, for example, in theory, be attractive to conceive of the need to at least maintain chemical, biological and even radiological programmes as a deterrent measure, but weapons only operate as a deterrent if there are reasons to think that they might be used. Furthermore, experience has taught us that we cannot trust Governments to stick to deterrence; that some of them cannot maintain the military strength to keep materials safe; and that, despite them maintaining programmes in a shroud of secrecy, there is only so much they can do in an era of ever increasing flow of people, goods and knowledge to prevent other states and even non-state actors developing, or obtaining and subsequently using as weapons, an ever-widening range of chemical, biological and radiological agents, many of which will have one or more legitimate, innocent uses to 'take cover' behind. Indeed, as earlier noted, state use continues to be a phenomenon in recent times. An additional threat of individuals and groups using agents to pursuing grudges or political objectives had also emerged strongly by the turn of the millennium (Tucker, 2000). If anything, it has become worse since, with some terror groups having the ability to operate in large geographical areas with near impunity for long periods and build up considerable resources. Proliferation is also a problem even when warfare use is discounted, bringing with it greater test and production related damage and threat to the environment, animals and human beings.

It is also important to address one final question about abusive warfare related projects and programmes of experimentation, namely whether or not we should be tempted to treat them as of largely academic relevance given that most of them which have been appear to have occurred a long time ago? The answer must be no for three reasons. The first is that they operate as an extreme warning to us about our capacity for inhumanity and our specific capacity to disastrously misuse medicine and in the process drawn on the human body in ways that fall dramatically short of ethical and legal norms of respect. The second is that such projects and programmes have 'lived on' long after they have officially ended. The 'energetic charge' of sentiment behind them is not fully dissipated, and their impacts continue to be felt in tangible ways. This has been made all the more so because of the way in which there occurrence has been 'handled.' As we have seen, for the most part, they were conducted secretively and then once found out about were not addressed in accordance to norms of moral and legal accountability and deterrence – i.e. swiftly, transparently, comprehensively and robustly. This speaks to a serious continuing shortfall in terms of respect. So too does our continuing use of data from the experiments for purposes beyond exposing and condemning them. Drawing from Bärnighausen's collation (2014, 92) of the literature concerning the ethics of such use, one may conclude that it:

- Treats the data as belonging to those using it when it is more correctly treated as belonging to those who were victimized to produce it and are victimized again each time it is used (with the user becoming a complicit torturer);

- Is the equivalent of building on top of Auschwitz (foregoing symbolic memorial);
- Implies those abused and killed were not human beings and keeps alive a philosophy of not valuing all human beings equally;
- Relativizes crimes against humanity, honours the perpetrators by implying good has come from their actions and signals to future scientists that immoral research is rewarded with citation and reference; and
- Corrupts the institution of medicine itself, signalling that it is concerned with pursuit of knowledge above ethics and humanity and also thereby undermining public confidence in it as a force for good.

These shortfalls also give rise to a third argument as to why we should not think abusive warfare related programmes and projects of experimentation are purely of academic relevance, namely that the ideological and practical conditions conductive to their existence have never been properly swept aside – meaning that the appearance that they are a thing of the past may, in some cases, be deceptive. In this field and others, we must exorcise the demons from our past both by completing remedial action and moving towards a new norm of describing, confronting and properly addressing atrocities and other shortfalls of respect. As Kleinman stresses, we inhabit a 'gritty, gray, everyday world of contradictory, untoward and inadequate moral realities' but must nonetheless resist inhumanity and advocate humanity (Kleinman, 2014, 1).

References

Advisory Committee on Human Radiation Experiments (ACHRE). (1995) *Final Report*. Washington, DC: US Government Printing Office.

Bärnighausen, T. (2014) Data generated in Japan's biowarfare experiments on human victims in China, 1932–1945, and the ethics of using them: In Nie, J-B., et al. (eds.) *Japan's Wartime Medical Atrocities*. Abingdon: Routledge, pp. 81–106.

Bashford, A. & Levine, P. (2010) *The Oxford Handbook of the History of Eugenics*. Oxford: Oxford University Press.

BBC News. (2004) Within prison walls Friday. *BBC*, 30 January. Available at news.bbc.co.uk/1/hi/programmes/this_world/3440771.stm.

BBC News. (2005) Japan rejects germ warfare claim. *BBC*, 19 July. Available at news.bbc.co.uk/1/hi/world/asia-pacific/4695151.stm.

BBC News. (2006) MI6 payouts over secret LSD tests. *BBC*, 24 February. Available at news.bbc.co.uk/1/hi/uk/4745748.stm.

BBC News. (2008) Porton Down victims awarded £3 million. *BBC*, 31 January. Available at news.bbc.co.uk/1/hi/england/wiltshire/7220225.stm.

Beecher, H. (1970) *Research and the Individual: Human Studies*. Boston, MA: Little, Brown, & Company.

Bernard, C. (1865) *An Introduction to the Study of Experimental Medicine*. New York, NY: Dover Press.

Childs, D. (2010) Doctor Hans Sewering: Physician whose activities under the Nazis caused controversy in later years. *The Independent*, 16 August.

CPS. (2011) CPS decides no prosecutions after re-review of evidence in Porton Down. *CPS*, 12 June. Available at cps.gov.uk/news/press_releases/132_06/index.html
Degen, P. (2014) Racial hygienist Otmar von Verscheur, the Confessing Church, and comparative reflections on postwar rehabilitation. In Nie, J-B., et al. (eds.) *Japan's Wartime Medical Atrocities.* Abingdon: Routledge, pp. 155–165.
Döring, O. (2014) Bioethics and exceptionalism: A German example of learning from medical atrocities. In Nie, J-B., et al. (eds.) *Japan's Wartime Medical Atrocities,* Abingdon: Routledge, pp. 141–154.
Ferster, E. (1966) Eliminating the unfit: Is sterilization the answer. *Ohio State Law Journal.* 27(4), 591–633.
Gareth Jones, D. & Whitaker, M. (2009) *Speaking for the Dead: Cadavers in Biology and Medicine.* Aldershot: Ashgate.
Gold, H. (1996) *Unit 731 Testimony.* Tokyo: Yenbooks.
Gomer, R., et al. (1981) Japan's biological weapons: 1930–1945. *The Bulletin of Atomic Scientists.* October, 43–53.
Greenfield, P. (2018) UK woman dies after being exposed to nerve agent Novichok. *The Guardian*, 8 July.
Grodin, M. (1992) Historical origins of the Nuremburg Code. In Annas, G. & Grodin, M. (eds.) *The Nazi Doctors and the Nuremburg Code: Human Rights in Human Experimentation.* Oxford: Oxford University Press, pp. 121–144.
Grunden, W. (2017) No retaliation in kind: Japanese chemical warfare policy in World War II. In Friedrich, B., et al. (eds.) *One Hundred Years of Chemical Warfare: Research, Deployment, Consequences.* Heidelberg: Springer Open, pp. 259–272. Available at https://link.springer.com/content/pdf/10.1007%2F978-3-319-51664-6.pdf.
Guillemin, J. (2005) *Biological Weapons: From the Invention of State-Sponsored Programs to Contemporary Bioterrorism.* New York, NY: Columbia University Press.
Guo, N. (2014) Taking responsibility for medical atrocities. In Nie, J-B., et al. (eds.) *Japan's Wartime Medical Atrocities.* Abingdon: Routledge, pp. 107–122.
Harris, S. (1999) Japanese medical atrocities in World War II: Unit 731 was not an isolated aberration. *International Citizens Forum on War Crimes and Redress,* Tokyo, Japan, 11 December. Available at www.vcn.bc.ca/alpha/speech/Harris.htm. Accessed 8.3.2019.
Harris, R. & Paxman, J. (1982) *A Higher From of Killing: The Secret History of Biological and Chemical Warfare.* London: Chatto & Windus.
Harris, R. & Paxman, J. (2002) *A Higher Form of Killing: The Secret History of Biological and Chemical Warfare.* London: Arrow Books.
Harvard Law School. (2018) *Nuremburg Trials Project.* Available at http://nuremberg.law.harvard.edu/nmt_1_intro. Accessed 8.4.2019.
Hammerstein, N. (1989). *Die Deutsche Forschungsgemeinschaft in der Weimarer Rupublik und im Dritten Reich.* Munich: Beck.
Holmes, O. & Phillips, T. (2017) Kim Jong-nam killed by VX nerve agent, say Malaysian police. *The Guardian*, 24 February. Available at www.theguardian.com/world/2017/feb/24/kim-jong-nam-north-korea-killed-chemical-weapon-nerve-agent-mass-destruction-malaysian-police. Accessed 8.3.2019.
Hornblum, A. (1997) They were cheap and available: Prisoners as research subjects in twentieth century America. *British Medical Journal.* 315, 1437–1441.
Hudson, C. (2007) Doctors of depravity. *Daily Mail*, 2 March.
Japan Times. (2000) Green Cross chiefs face prison term for HIV scandal. Available at www.japantimes.co.jp/news/2000/02/24/national/green-cross-chiefs-receive-prison-terms-for-hiv-scandal/#.XGutaOj7QuU. Accessed 8.3.2019.

Jones, T. & Simkin, M. (2002) Tokyo court rejects germ warfare compo claims. *Australian Broadcasting Corporation LATELINE*, 28 August.

(The) Khabarovsk Trial. (1950) *Materials of the Trial of Former Servicemen of the Japanese Army Charged with Manufacturing and Employing Bacteriological Weapons.* Moscow: Foreign Language Publishing House.

Kleinman, A. (2014) Introduction: Medical atrocity, history and ethics. In Nie, J-B., et al. (eds.) *Japan's Wartime Medical Atrocities.* Abingdon: Routledge, pp. 1–19.

Kolb, S., et al. (2012) Apologizing for Nazi medicine: A constructive starting point. *The Lancet.* 380, 722–723.

Kuhl, S. (2002) *The Nazi Connection: Eugenics, American Racism and Nazi Socialism.* Oxford: Oxford University Press.

MacDonald, D. (2014) The U.S.'s memory problems: Diaspora groups, civil society and the perils of 'chosen amnesia'. In Nie, J-B., et al. (eds.) *Japan's Wartime Medical Atrocities.* Abingdon: Routledge, pp. 166–182.

Matsubara, Y. (1998) The enactment of Japan's sterilization laws in the 1940s: A prelude to postwar eugenic policy. *Historia Scientiarum.* 8(2), 187–201.

McCurry, J. (2011) Japan unearths site linked to human experiments. *The Guardian*, 21 February.

McCurry, J. (2018) Unit 731: Japan discloses details of notorious chemical warfare division. *The Guardian*, 18 April.

Miazga, L., et al. (1974) Interview of Shields Warrens, (transcript of audio recording), 9 April. Available as Document No. DOE-121294-D-14 in Advisory Committee on Human Radiation Experiments. (1995) *Final Report.* Washington, DC: US Government Printing Office.

Miller, V., et al. (2004) *Health Effects of Project Shad Biological Agent: Bacillus Globigii.* Report Prepared for the National Academies of the Institute of Medicine by The Center for Research Information, Inc. Available at www.nationalacademies.org/hmd/~/media/Files/Report%20Files/2007/Long-Term-Health-Effects-of-Participation-in-Project-SHAD-Shipboard-Hazard-and-Defense/BACILLUS-GLOBIGII.pdf. Accessed 8.3.2019.

Mlodoch, K. (2017) The indelible smell of apples: Poison gas survivors in Halabja, Kurdistan-Iraq, and their struggle for recognition. In Friedrich, B., et al. (eds.) *One Hundred Years of Chemical Warfare: Research, Deployment, Consequences.* Heidelberg: Springer Open, pp. 349–362. Available at https://link.springer.com/content/pdf/10.1007%2F978-3-319-51664-6.pdf.

Nichibenren (Nihon Bengoshi Rengokai, Japan Federation of Bar Association). (1968) *Jinken Hakusho* (Human Rights Report). Japan: Nihon Bengoshi Rengokai.

Nickson, J. (1951) Blood changes in human beings following total-body irradiation. In Stone, R. (ed.) *Industrial Medicine on the Plutonium Project: Survey and Collected Papers.* New York, NY: McGraw Hill Book Co.

Nie, J-B. (2004) The West's dismissal of the Khabarovsk trial: Ideology, evidence and international bioethics. *Journal of Bioethics Inquiry.* 1, 32–42.

Nie, J-B. (2014) The US cover-up and China's responses. In Nie, J-B., et al. (eds.) *Japan's Wartime Medical Atrocities.* Abingdon: Routledge, pp. 123–138.

Nish, I. (2000) Nationalism in Japan. In Liefer, M. (ed.) *Asian Nationalism.* Abingdon: Routledge, pp. 82–90.

O'Flynn, E. (2015) US bomber crew shot down over Japan were dissected while ALIVE in horrific WW2 experiments: Japanese university acknowledges full details of atrocity 70 years on. *MailOnline.* 7 April. Available at www.dailymail.co.uk/news/

article-3028694/U-S-POWs-shot-Japan-70-years-ago-dissected-ALIVE-macabre-experiments-controversial-new-exhibition-shows.html. Accessed 8.3.2019.

Osler, W. (1907) *The Evolution of the Idea of Experiment in Medicine.* Toronto: The Hannah Institute for the History of Medicine.

Otsubo, S. (2005) Between two worlds: Yamanouchi Shigeo and Eugenics in early twentieth-century Japan. *Annals of Science.* 62(2), 205–231.

Ploetz, A. (1895) *Grundlinien einer Rassen-Hygiene.* Berlin: S. Fischer.

Pollock, A. (1997) Japan blood supplier, facing H.I.V. penalty, to be acquired. *The New York Times*, 25 February, p. 7 (section D, column 1).

Proctor, R. (1992) Nazi doctors, racial medicine and human experimentation. In Annas, G. & Grodin, M. (eds.) *The Nazi Doctors and the Nuremburg Code: Human Rights in Human Experimentation.* Oxford: Oxford University Press, pp. 17–31.

Ralstin-Lewis, D. (2005) The continuing struggle against genocide: Indigenous women's reproductive rights. *Wicazo Sa Review.* 20(1), 71–95.

Reich, W. (2013) *The Mass Psychology of Fascism* (3rd Edition). New York, NY: Farrar, Strauss & Giroux.

Riedel, S. (2004) Biological warfare and bioterrorism: A historical review. *Proceedings (Baylor University Medical Center).* 17(4), 400–406.

Roberts, A. (2008) Medal for murder: Germany honours Nazi SS doctor linked to death of 900 children in death camps. *The Daily Mail*, 28 May.

Rothman, D. (1998) The Nuremburg Code in the light of previous principles and practices in human experimentation. In Trohler, U., et al. (eds.) *Ethics Codes in Medicine: Foundations and Achievements of Codification Since 1947.* Altershot: Ashgate, pp. 50–59.

Ryall, S. (2017) Inside North Korea's barbaric prisons where inmates are starved, tortured, undergo forced abortions and dig their own graves. *The Mirror*, 21 June. Available at www.mirror.co.uk/news/real-life-stories/north-korea-prisons-otto-warmier-10653494. Accessed 8.3.2019.

Sass, H.-M. (2005) Emergency management in public health ethics: Triage, epidemics, biomedical terror and warfare. *Eubios Journal of Asian and International Bioethics.* 15 September, 161–166.

Schmaltz, F. (2017) Chemical weapons research on soldiers and concentration camp inmates in Nazi Germany. In Friedrich, B., et al. (eds.) *One Hundred Years of Chemical Warfare: Research, Deployment, Consequences.* Heidelberg: Springer Open, pp. 229–258. Available at https://link.springer.com/content/pdf/10.1007%2F978-3-319-51664-6.pdf.

Schmidt, U. (2015) *Secret Science.* Oxford: Oxford University Press.

Schoppa, R. (2011) *In a Sea of Bitterness; Refugees in the Sino-Japanese War.* Cambridge, MA; and London: Harvard University Press.

Shimizu, T. (1979) *Seitai Jikken* (Experimentation with Living Human Subjects), expanded version. San-Ichi Shobo.

Spratt, B. (1999) *Independent Review of the Possible Health Hazards of the Large – Scale Release of Bacteria During the Dorset Defence Trial.* London: Ministry of Defence.

Syal, R. (2002) Porton Down used soldiers for Sarin gas tests in 1983. *Sunday Telegraph*, 13 October.

Takasugi, S. (1973) *Nihon Iryo no Genzai* (The Original Sin of Japanese Medicine). Aki Shobo.

Task Force on Human Subject Research. (1994) *A Report on the Use of Radioactive Materials in Human Subject Research that Involved Residents of State-Operated Facilities Within the Commonwealth of Massachusetts from 1943 to 1973*, April.

Thompson, M. (2002) 'Millions exposed' in 1960s experiments. BBC News, 13 September. Available at news.bbc.co.uk/1/hi/england/2254664.stm

Trapp, R. (2017) The use of chemical weapons in Syria: Implications and consequences. In Friedrich, B., et al. (eds) *One Hundred Years of Chemical Warfare: Research, Deployment, Consequences.* Heidelberg: Springer Open, pp. 363–375. Available at https://link.springer.com/content/pdf/10.1007%2F978-3-319-51664-6.pdf.

Tsuchiya, T. (2003) In the shadow of the past atrocities: Research ethics with human subjects in contemporary Japan. *Eubios Journal of Asian and International Bioethics.* 13, 100–102.

Tsuneishi, K. (2014) Unit 731 and the Japanese Imperial Army's biological warfare programme. In Nie, J.-B., et al. (eds.) *Japan's Wartime Medical Atrocities.* Abingdon: Routledge, pp. 23–31.

Tucker, J. (ed.) (2000) *Toxic Terror: Assessing Terrorist Use of Chemical and Biological Weapons.* Belfer Center Studies in International Security. Cambridge, MA; and London: MIT Press.

Vollmann, J. & Winau, R. (1996) Informed consent in human experimentation before the Nuremberg code. *BMJ.* 313, 1445–1449.

Wang, S. (2014) War crimes trials and post-war politics. In Nie, J-B. (ed.) *Japan's Wartime Medical Atrocities.* Abingdon: Routledge, pp. 32–58.

Whittemore, G. (1995) Shields Warren Papers: A Cumulative Update of Excerpts. ACHRE No. BU-030395-A-1, 3rd March in Advisory Committee on Human Radiation Experiments. *Final Report.* Washington, DC: US Government Printing Office.

Williams, P. & Wallace, D. (1989) *Unit 731: The Japanese Army's Secret of Secrets.* London: Hodder & Stoughton.

Withington, C. (1886) *The Relation of Hospitals to Medical Education.* Boston, MA: Cupples, Upham & Company.

The Working Group on Human Subject Research to Philip Campbell. (1994) *The Thyroid Studies: A Follow-up Report on the Use of Radioactive Materials in Human Subject Research that Involved Residents of State-Operated Facilities Within the Commonwealth of Massachusetts from 1943 through 1973*, June.

Wright, M. (2018) Dark past of Porton Down: The secret nerve agent tests that led to the death of a serviceman. *The Telegraph*, 10 March.

Wyzanski, C. (1946) Nuremburg: A fair trial? A dangerous precedent. *The Atlantic.* April, 1–4.

Yudin, B. (2014) Research on humans at the Khabarovsk Trial. In Nie, J-B., et al. (eds.) *Japan's Wartime Medical Atrocities.* Abingdon: Routledge, pp. 59–78.

5 Solid organ transplantation

5.1. Introduction

In just over a century the practice of transplantation has emerged from being a highly speculative and marginal form of medical practice to being practiced at mass scale in many forms. Historical analyses tend to highlight the fact that this transition could not have occurred without a series of technological developments. However, it is equally true that it could not have occurred without the extensive support from key stakeholders. States and medical establishments have facilitated its practice and, alongside many individuals professionals, propagated the idea that it is a public good. The public has largely bought into this idea, many even being willing to receive and/or donate materials for it. A halo has resultantly surrounded transplantation as a phenomenon. Neither its practice or the way materials have been procured for it have been immune from serious criticism, but for the most part, critiques of it have focused on specific issues and tended to be made without the backing of a properly justified theoretical standpoint. This chapter breaks the mould in both regards by building a compound analysis of the key problems that relate to both from the theoretical perspective of respect for worth developed and justified in Chapter 1. Its main body falls into three parts in this regards: firstly, a better exposure than yet delivered of the great extent to which transplantation has been built on experimental excess; secondly, a more complete analysis than hitherto provided of the more fundamental issues that arise with transplantation as a form of medicine with an analysis of the extent to which these raise questions about the extent it is relied on and, in some senses, on whether it should be relied on at all; and thirdly, an analysis of the full extent to which the ways in which procurement has been approached in discourse, practice, governance and even law are not only problematic but also further weaken the case for relying on transplantation.

5.2. Transplant as a story steeped in experimental excess?

Part of the bedrock of clinical development of transplantation as a therapy lies in grafting for plastic surgery, which started around 1000 BC with the Indian surgeon Sushruta using skin flaps to replace cut off noses (Champaneria et al.,

2014). The first references to the performance of transplantation itself start six centuries later in Chinese texts. They include reference to the eminent Chinese surgeon, Tsin Yue-Jin exchanging the hearts of two soldiers; one endowed with a strong will but a weak spirit and the other conversely afflicted (Crawford et al., 2013). In the third century AD, Cosmas and Damian, the patron saints of doctors, are reputed to have performed transplants, including a leg transplant (Dewhurst, 1988) and were influential in setting a tone of body parts as both sacred objects and commodity which swept through Europe in the middle ages (Decker, 2014, XIV).

Further reports of use in Europe occur through the Renaissance period (Deschamps, 2005, 2–3). However, for practical purposes, the meaningful start to clinical use of transplantation is best put at around 1860 when the French neurologist Charles Édouard Brown-Séquard (1817–1894) started to inject semen, testicular blood and gonadal extracts from animals into human beings (Aminoff, 2017). The experiments were expensive, useless, dangerous and founded on harm to animals. They were followed not long after by attempts at thyroid transplant, which persisted until near the end of the century in spite of the fact that they had poor results in terms of graft survival and led, in many cases, to recipient death (Decker, 2014, 9).

Better clinical outcomes were gained in the early 1900s with blood and cornea (Zirm, 1906; Brightbill, 2009, 2), but the theme of experimental excess continued to dominate. Kidney transplantation was tried many times over in the first half of the twentieth century on little more than the back of the fact that Caroll developed the surgical ability to join blood vessels at the end of the nineteenth century (Townsend et al., 2016, 617). In the first decade, these transplants were carried out extensively and persistently between animals in spite of the fact that the first few efforts alone were sufficient to make it clear that results in terms of graft and/or subject survival were going to be extremely poor unless or until some major breakthrough in understanding of graft rejection was achieved. They were also progressed into the human domain with equally poor results. This started in 1902 with Imre Ullmann (1861–1937) conducting the first animal-human kidney transplant between a pig and a young women with serious uraemia. The graft was acutely rejected, and the patient died (Hatzinger et al., 2016).

After this experience, Ullmann came to the conclusion that he should cease his work on ethical grounds. However, several other clinicians were still determined to have their own piece of the experimental action. Having already connected human kidneys to the arms of two uraemic patients, the Frenchman Mathieu Jaboulay (1860–1913), transplanted a pig kidney into one human patient and a goat kidney into another in 1906 (Reemtsma, 1995). Then in December 1909 Ernst Unger conducted a kidney transplant from a monkey to a human, having transplanted the kidney of a still-born child into a monkey earlier in the same month. He continued to perform further experiments in this vein (Bos, 2017, 16) in spite of concluding from the poor results that there was a biochemical barrier to transplantation (Unger, 1909). Taken as a whole, the animal-to-human kidney transplants performed in this era clearly did not meet standards of respect for human worth. Bos has suggested that they were performed on human beings as a last resort (2017, 17), but a better description of them is that they were performed opportunistically on unconscious people whose relatives had been exploited into giving permission to an experiment that was incompatible with their worth because it had no real prospect of generating the kind of significant benefit that would 'ethically offset' its intrusiveness.

World War I and its legacy of needing to help the injured and rebuild institutional capacity temporarily put a stop to these experiments. However, they returned later partly through the work of the Ukrainian surgeon Yurii Voronoy (1895–1961). Voronoy not only conducted more kidney transplant experiments between animals but went on in 1933 to carry out a series of six experiments between humans. The only positive that came out of these experiments was that two of the recipients were able to mount some recovery of their own function in the short period for which their grafts lasted (Bos, 2017, 16). Whilst there were several other attempts at human-human kidney transplant in the 1940s and early 1950s, it was not until 1954 in Boston, US that the first long term graft survival was achieved (Murray et al., 1955).

Medawar had already established in 1944 that graft rejection was an immunological phenomenon (Medawar, 1944). The immune system naturally fights against the presence of material it perceives to be incompatible with the body and thus long term graft survival of solid organs and tissues is only achievable in cases where the immune response is weaker. In the case of corneal transplant, it is natural for it to be weak because of 'anatomical, physiological, and immunoregulatory factors' particular to the cornea (Niederkorn, 2014). However, for other solid organs and tissues, it is natural for it to be strong and weakness has to be 'manufactured.' What will suffice to enable long term graft survival will depend heavily on the type of material that is being transplanted. For kidneys, outside of cases involving exceptional match (such as where material is from an identical twin as was the case in Boston), measures have to be taken to suppress the immune response. Early efforts at immunosuppression were coarse to the point of counterproductivity. For example, the use of total body irradiation caused severe damage and even death in some cases (though it has subsequently been used with some success in the transplantation of stem cells).

The real breakthrough did not come until 1962 with the advent of corticosteroid use. The development of cyclosporine built on this. It was used in clinical trials on patients from 1978 and by the mid-1980s had come to be routinely used on its own or – more commonly – in combination with other immunosuppressants including corticosteroids. Its introduction had a pronounced impact on graft survival and morbidity/mortality rates. For example, it was central to 1-year kidney graft survival rates rising from 50% to 80% in the 1980s and mortality rates decreasing from 10–20% to minimal (Muntean & Lucan, 2013).

The transplant community treated its availability as a cue to expand transplantation – not least of kidney, lung, heart and liver (Ballantyne et al., 1989). However, even through a lens of overall support for transplantation, it is clear that this expansion had downsides. One of these was its heavy reliance on cadaveric procurement which in turn was facilitated by approaches to defining and testing for death that were controversial then and remain so now (see Section 5.4). A second was a lack of precaution around cyclosporine use. Many early medical reports simply giddily assigned it wonder drug status (Fox & Swazey, 2017, 3). By the end of the 1980s, it had become all too painfully evident that a range of serious side effects could result from its use, including acute and chronic nephrotoxicity, lymphomas, hypertension, hepatoxicity, hirsutism, anaemia, neurotoxicity, endocrine and neurological complications and gastrointestinal distress (see further Tedesco & Haragsim, 2012). A third issue was that some forms of transplantation only achieved reasonable clinical results in this period, having left a trail of experimental destruction in their wake. The history of heart transplantation particularly exemplifies this. In the wake of Barnard's two initial heart transplants in 1967, '(m)uch damage was done to the image of heart transplantation by the immediate unseemly scramble to get on the bandwagon' (Hoffenburg, 2001, 1479). In the following year, 107 such transplants were carried out by 64 surgical teams in 24 countries with predictably bad results:

> operations were performed by ill-trained surgeons without proper back-up, matching of donors and recipients was poor, there was little appreciation of the need for meticulous aftercare and the management of rejection.
>
> (Hoffenburg, 2001, 1479)

The sense in which the patient was little more than an afterthought was added to by the celebrity atmosphere that developed around this pioneering work. Surgeons were hyped in the media and some of them were intent on milking it. Barnard went on a global lap of honour, and Donald Ross, having performed a copycat procedure, was photographed for newspapers bearing aloft a Union Jack and a poster saying 'We're backing Britain' – the unseemliness of which was made all the more stark by the patient dying shortly after (Hoffenburg, 2001). Another example of experimental destruction was animal-human solid organ transplants. These transplants killed primates and other mammals and, where used as a therapy in their own right rather than as a bridge, did little for the desperate patients who relied on them because grafts lasted at best a matter of months and more commonly a matter of hours or days. There use withered by the 1980s – though not so much out of recognition of the aforementioned problems as the simple fact that human-human solid organ transplant had been dramatically expanded and was now providing decent graft survival rates. Nonetheless, xenotransplantation with respect to solid organs has been persisted with on a significant scale experimentally in an even more radical form consisting of primates having organs transplanted into them from especially created animal-human hybrids, such as pigs modified with human genetic materials (for a review of studies, see Cooper, 2014). The development of this field continues to be touted as one way in which demand for transplantation of solid organs into human beings could be satiated. However, there are a range of serious objections to it, including that it: will likely be very costly bearing in mind, not least, the desire of developers to commercially profit from it; will transgress the integrity of natural forms (see next section); is abusive of the Chimeras created for it; *and* brings with it the potential for cross species transmission of viruses, including even the potential for causing a serious pandemic (Butler, 1998).

The modern era has also been characterized by other forms of experimental excess. One issue in this regard concerns the expansion of categories of person who are used as donor – with some ending up being included when on health grounds they really ought to be excluded, such as where they are prehypertensive or obese (Thukral et al., 2018). Another issue concerns expansion into new and highly questionable forms of transplantation. One of these is stem cell infused trachea (windpipe) transplant. The basic form of is technique involves taking a donor windpipe, stripping it of the donors cell and seeding it with the patient's own stem cells. It was developed by the Swiss born Italian surgeon, Paolo Macchiarini, who has subsequently been accused of research misconduct and performing unethical experimental surgeries. Whilst the technique was used ten times up to 2014, eight of the patients have already died. It seems that the stem cells do not survive and the only successes have come with patients who have had a stent. Stents mitigate the risk of the transplanted material collapsing. A recent investigation by BBC Newsnight has highlighted the problem of the technique recently being used with a teenage patient by a University College London (UCL) group led by Professor Burchill (Newsnight, 2019). They added new dimensions to the technique including freeze thawing which made it more likely that the transplanted windpipe would collapse. There are concerns that the patient's mother was not properly informed of the poor results of most of the transplants conducted prior to hers or of the particular added risk of collapse posed by freeze-thawing and lack of a stent. It seems that the UCL team, which has now discontinued its work, prioritized further developing the field over performing the intervention in a way that tried to contain the already radically high level of risk posed by it. The over-optimism it had with regards to its work extended to a disturbingly cavalier treatment of the facts with respect to the teenagers' death – seven documents produced by the team in its aftermath are all misrepresentative in one or other significant way (Newsnight, 2019).

Transplantation is also increasingly being resorted to in circumstances where it has no life protecting function and thus brings reduced life expectancy (of about 10 years on average) due to the fact that recipients will be on a toxic immunosuppressive regime. Composite tissue forms of transplantation – such as transplants of face, hand, part of knee and so on – are normally a case in point. Finally, outright abusive practice continues to occur. A case in point is the work of the Italian neurosurgeon, Dr Sergio Canavero on full body transplants (sometimes more questionably described as head transplants). He has already performed such transplants between animals which even from a non-fundamentalist animal welfare perspective is insidious considering that they can never form a morally acceptable platform from which execute his plan to transplant a human body onto the head of a living human being (Dauber, 2017).

5.3. More fundamental issues with transplantation as a form of medicine

The idea that gifting body material to others is prima facie a good thing is a typical stance taken by those who write about the relationship between spiritual thought and transplantation. However, this stance is open to criticism on a number of levels. Firstly, whilst it is reasonable to associate spirituality with compassion and compassion with need, it should not be assumed that what human beings actually ultimately need is precisely the same thing as conventional norms suggest. Consider the following anthroposophical reflection on the appropriateness of becoming a recipient of an organ:

> What can be said about the situation of the recipient? The first question to ask oneself is what would be the significance of the destructive restructuring of an organ to the point where it is no longer able to perform its function in the medical sense and has to be replaced if the patient is to go on living? Aspects of this are discussed in two lectures by Rudolf Steiner . . . where he speaks of organs being destroyed for karmic reasons 'by the benevolent powers', giving the person concerned a chance to deal with the causes of the disease and thus creating an opportunity for further development. Transplantation intervenes in this process of interaction between the human being and the hierarchies. Transplantation does not discover and deal with the cause of pathologic organ destruction as is evident, for instance, from the fact that the renal disease may also manifest in the transplant. . . .
>
> Patients must of necessity be immunosuppressed. The immune system enables us to maintain our boundaries against the environment and exist as individual entities. It may thus be seen as a function of the ego organization. Medication prevents the ego organization from intervening at this level. The question as to how far this also disrupts or inhibits the ego organization in other areas will be left open for the moment.
>
> A transplanted organ is also a denervated organ. The body-bound astral body is no longer able to perform its regulatory function via the autonomic nervous system.
>
> (Baumgart, 1994)

These sentiments speak to a holistic understanding of reality that is common to traditional systems of medicine across the world and the wider spiritual traditions within which they 'sit.' They also fit comfortably with the ethos of the Great Eastern Religions. From these perspectives, there is also another reason to question the sense of donating body materials which is that are antagonistic to the spiritual purpose for which they exist.

Transplantation may also be described as problematic from the slightly different and more specifically bioethics based perspective that it interferes with human nature and form. The modification it entails is fundamental rather than minor since it artificially 'stitches' to one human being a part or parts from another or even from an animal. This creates what is technically known as a Chimera. Chimeras have long been part of our mythology, most often conceived and depicted in an inter-species form. As Deschamps et al., note:

> Pre-historic cave paintings rarely represent people but, in the Lascaux cave in France (ca 15000 BC), the only representation of a human being shows a man with a head of a bird. . . . The gods of Ancient Egypt are often represented with the body of a man and the head of an animal: Anubis has the head of a jackal; the Great Sphinx of Giza (ca 2500 BC) is a lion with the head of a woman.
>
> (2005, 91)

Chimeras were a significant part of the mythology of ancient Greek culture – being referred to, for example, in Homer's Iliad. There is also some ancient mythology that describes them being specifically generated through transplantation. For example, twelfth century BC Sanskrit texts refer to Ganesha receiving the head of an elephant after being beheaded by Shiva (Bhandari, 1997). However, the idea of an uncreated creator creating such creatures is entirely different from humans doing so. The idea that the latter is a hubristic misuse of human will and brings an abundance of disastrous consequences is reflected notably in the myth of Atlantis and Mary Shelley's novel *Frankenstein.* That the warnings with respect to transplantation are not more readily recognized suggests we may have collectively largely forgotten how radically altering it is as a technique. We may also be some way through a shift in values that has undermined our ability to both mount an impulsive 'yuk factor' response (Midgley, 2000) and to give credence to the idea that this response may in some cases relate to important values.

Even if one rejects these critiques, there are still reasons to think that transplantation is a misanthropic use of resources. Within most well-resourced health systems, it has become normal to view a wide range of forms of it as cost effective and practice them on an extensive basis. For solid organ transplantation specifically, a Quality Adjusted Life Year (QALY) case has been made for continued and expanded use of kidney and liver transplant (see especially Jarl & Gerdtham, 2011) and to some extent even for lungs and hearts (e.g. Tiessen et al., 2008; Jarl & Gerdtham, 2011). However, the figures are not startling. For example, over dialysis, kidney transplant was assessed in 2003 as only bringing 2 years extra QALY and 3.6 years extra life expectancy in the case of 'cadaveric donors' being used and 5.8 years extra QALY and 10.5 years extra life expectancy when living donors were used (Knoll & Nichol, 2003). Furthermore, a QALY based analysis gives a very partial picture. It cannot really tell us whether transplantation is an appropriate basis for dealing with the problems that underlie the claimed health need for it which is certainly a valid question given that it embodies and encourages the kind of mechanistic and industrialized approach to medicine that I described at the end of Chapter 1 as outmoded and counterproductive. Nor can it tell us whether we might be better off instead devoting attention and resource on better understanding the cause of the problems that give rise to the claimed need for it and where appropriate taking preventative and curative action.

What we do know on this front is that the kinds of solid organ problem specifically that give rise to the claimed need for transplantation are associated in about 50% of cases with one or more environmental and lifestyle factors (Levitt, 2011). The most implicated factors include pollution, inadequate housing and homelessness to poor diet, lack of exercise, use of tobacco and misuse of drugs and alcohol – are causally significant not just for many health problems in general but in around half of the cases of organs failing. Self-evidently these issues exist because of human behaviour and can be resolved by changes in it.

As a case study, Type 2 diabetes provides one good example of the need and potential for such change. Globally, it is a very serious and growing health issue that is a major cause of the high and growing global incidence of kidney failure (Wetmore & Collins, 2016) which now affects over 2 million people (Eggers, 2011). Treating the symptoms of Type 2 is becoming very costly. In England, for example, it takes up 10% of the National Health Service budget (Boseley, 2018). However, there is evidence to suggest that at the population level a slight reduction in weight can significantly reduce its burden (Franco et al., 2013). Furthermore, a growing body of evidence suggests that amongst those who are particularly susceptible to getting it, lack of exercise and poor diet have a particularly pronounced negative impact (Davies, 2003), and conversely, even small increases in physical activity and decreases in calorific intake significantly reduce risk (Knowler et al., 2002; Tuomilehto, 2001; Ramachandran, 2006; Pan, 1997). A recent review has suggested that management of it ought to routinely include management of diet (Sami et al., 2017). Results from an ongoing clinical trial in Newcastle-Upon-Tyne, England, are so far suggesting that remission of it can be brought about by maintaining a low calorie diet (DiRECT, 2018), and this proposition is currently being further tested across England in a trial with 5,000 Type 2 diabetes patients (NHS England, 2018).

In the light of the potential for prevention alone, the transplant centric focus of many countries is open to question as is the widely perpetuated assumption that the gap between supply and demand of transplantable organs reflects an insufficiency of supply rather than partly or wholly being attributable to an excess of demand.

5.4. The procurement of body materials for transplantation

Death and procurement

As death ends interests related to body functioning, it opens up the range of forms of intrusion on the body that are capable of being justified even within the confines of a respect for worth based framework of ethics. For example, burying, burning or extracting essential to life organs are inconsistent with human worth when done to the living but not always so when done to the dead. This is one of the more obvious reasons why it is important to arrive at an accurate or otherwise precautionary assessment of what death is and also to ensure that we use an accurate or otherwise precautionary approach to determining when it has occurred. In other words, we need to be conservative in our approach to definition, criteria and tests.

The first issue of controversy has been over whether death is a process or an event. The online etymology dictionary notes how the old English word *deað* is derived from Proto-Germanic *dauthuz* which is also the source of related variants such as the Old Saxon *doth*, Old Frisian *dath*, Dutch *dood*, Old High German *tod*, German *Tod*, Old Norse *dauði*, Danish *død* and Swedish *död*. Meanwhile the Oxford English Dictionary defines death as 'the action or fact of dying or being killed; the end of the life of a person or organism, the permanent ending of vital processes in a cell or tissue' (OUP, 2019). Unfortunately, this is not a specific definition of death but rather a dragnet of several distinct and ultimately incompatible definitions. As has been noted and described elsewhere, there is a process of loss of function that leads up to death (Miles, 1999, 313; Price, 2000, 50), but this is clearly not the same as saying that death is a process. Indeed, even if we reached a point where it became semantically or clinically normal to view it as a process, it would never be right to treat it as such ethically or legally since the differences between the interests and related rights of the living and the dead make it imperative to be clear which state one is in at any given point. Luckily, the idea that it is an event also fits with the common sense that the two states are mutually exclusive, meaning that at any one moment a person must be alive or dead (see further Bernat, 2002; cf. Halevy & Brody, 1993; Emanuel, 1995).

The next issue concerns what characterizes it as a state. If one thinks of life as pervading the whole organism, the criteria for death would need to be based around a total loss of function. Somewhat in this vein, the ancient Greeks and Romans did not bury human beings until after putrefaction had set in. By contrast in modern times there has been a marked shift towards using irreversible loss of function of the brain as a criterion for determining when death has occurred and in some cases treating irreversible loss of circulation and respiration or even irreversible loss of circulation alone as a basis for intimating that a criterion of irreversible loss of brain function had been met, or as a criterion for determining death in its own right.

The reference to these three types of function is significant since it is commonly clinically considered that they work as an interdependent triad such that loss of one will lead to loss of the others and in turn ultimately lead to loss of function of the organism as a whole (Pallis & Harley, 2009). However, that is not enough to treat them as having a privileged place within the determination of death, since there are also a number of other body functions which, if lost, will inexorably lead to loss of function in the organism as a whole It would happen, for example, if one totally lost function of one's kidneys or liver or adrenal glands or intestines (Accad, 2015, 220). To assert that loss of function of a specific vital body part means that death has occurred, rather than simply that a person is dying, requires us to embrace the idea that presence of the function concerned is somehow essential to what it means to be a living human being.

In that respect, it may be noted that at least when one looks across different spiritual forms of thought there generally is a privileged status afforded to one or more of the brain, heart and lungs and/or their functions that enjoy a privileged status. Furthermore, of the three, one can point to the brain being especially significant because it has an association with our awareness and acts as a kind of central processor with respect to the functions of our bodies as a whole. Nonetheless, I do not think that there is an authoritative basis on which to attempt to determine death until function of the whole organism is lost. In spiritual terms, our essence is immaterial and unbounded in its relationship to the body. Death may reasonably be described as some kind of fundamental change between it and the body, but precisely what that change is and how it is represented in bodily processes is not exoterically clear.

There are also several specific problems with modern approaches to the determination of death. One which relates particularly to brain-based approaches concerns the way in which they were brought into being and were able to gain and maintain a dominant hold on law and practice. It started with ventilated patients, specifically, the American neurologist Robert Schwab used neurological tests as a basis for declaring such a patient dead in 1954 (Belkin, 2003), and the French neurologists Wertheimer and Jouvet did the same with another such patient in 1959 (Wertheimer & Jouvet, 1959). Wertheimer and Jouvet characterized the patient as having had death of the nervous system, but that same year, two other French neurologists, Mollaret and Goulon, when faced with a patient in the same clinical state, concluded that they were should not be defined as dead but rather as in a state of irreversible coma, which they termed *coma dépassé* (Mollaret & Goulon, 1959). Seemingly unaware of this work, the Ad Hoc Committee of the Harvard Medical School to Examine the Definition of Brain Death ended up titling the influential 1968 report that resulted from its deliberations *A Definition of Irreversible Coma* and writing up its content in such a way as to make clear that it regarded irreversible coma as equating to death.

So what impelled clinicians to propose a state of irreversible loss of brain function without proof of its existence? And, more pertinently, what impelled some of them to equate such loss with actual death since '(n)o amount of neurological study could possibly determine whether those with dead brains should be considered dead people' (Veatch, 1999, 140). The answer appears to have been that doing so had instrumental value. As we have seen it has filled the role of enabling some patients on ventilation to be deemed to be dead. This has enabled ventilatory support to be removed from them with less ethical and legal complication and potential objection than if they were still deemed to be alive. It has also opened up 'a new frontier' in the procurement of organs for transplantation (Evans & Hill, 1989, 114). This also started happening prior to the Harvard report coming out. Indeed, in 1963, the Belgian surgeon Alexandre defined a patient in *coma depasse* as dead in order to effect an otherwise unjustifiable removal of that patient's kidney for transplantation (Machado, 2005). Even more extreme than this, it has gone on to be routinely used to allow even vital to life materials to be removed from those who would otherwise have been classed as alive.

Aside from being physicalistic and opportunistic, brain-based approaches draw on criteria for measuring death which can be satisfied without the brains capacity for integrated functioning being entirely lost. Patient reports illustrating as much started to emerge in the late 1980s and were collected together and published by the renowned neurologist Alan Shewmon (1998) who withdraw his prior support for the concept of brain death in the light of them. By the early 1990s, Halevy and Brody (1993) had concluded that many patients who fully meet any of the standard criteria for measuring the death of the brain, in fact, retain not just various residual functions in the body but also considerable, measurable brain function. Shah has summed up the situation by noting that patients who meet brain death criteria can still produce electroencephalographic activity and auditory and/or visual evoked potentials and can perform functions ranging from 'circulating blood, maintaining respiration and body temperature, regulating salt and water homoeostasis, digesting food, healing wounds, fighting infections, mounting a stress response to surgical incisions, maturing and growing proportionally (in children) and even gestating foetuses successfully' (Shah et al., 2011, 720).

One possible response to this problem is to maintain a brain-based approach to death but move the goal posts away from requiring loss of integrational power to loss of something else. The President's Council on Bioethics 2008 white paper 'Controversies in the Determination of Death' suggested loss of ability to engage in the essential work that defines living things (President's Council on Bioethics, 2008). However, in envisaging such work to include breathing and consciousness the Council was unwittingly providing a basis for the tentacles of brain death to be extended to encapsulate various additional categories of the human species that have traditionally been viewed as alive – including foetuses, those in an apnoeic coma and high cervical quadriplegics in a vegetative state.

The only basis on which one could now realistically try to use current testing regimes as a basis for determining brain death would be if one put aside the idea of a whole brain and even a brain stem based approach and aimed for something much more selective. One example of such selectivity is the attempt to equate death with irreversible loss of all capacity for consciousness. However, I would describe consciousness as awareness that is incapable of being lost as it is part of one's soul, not a function that is reliant on the operation of one's brain. As such I would argue that so-called difficulties of consciousness should just be treated as difficulties that the soul has in working via the brain because of certain damage to it. It is not clear how treating such difficulties as a basis on which to define one as dead could be described as anything other than personism (cf. Veatch, 2005).

A final problem relates to the consequences of the fact that every set of criteria for diagnosing so-called brain death includes a test to determine whether breathing is absent (an apnoea test). Where a patient is ventilated this means that their ventilation is switched off for up to 10 minutes to see if there are indicators of cardiac arrest having occurred in the form of carbon dioxide increases in the blood and a drop in blood pressure. This test significantly impairs the possibility of recovery and can lead to the death of the patient through a heart attack or irreversible brain damage (Cuo, 2006). It is regularly performed without the involvement (let alone agreement) of kin.

As I noted in the introduction to this section, some jurisdictions do not confine themselves to a brain-based approach to death. For example, the majority of states in the US, model their local laws on section 1 of the US Uniform Determination of Death Act 1980 which stresses that death means the irreversible loss of capacity of the organism for integrative functioning, but can be determined as having occurred both where the whole of the function of the brain has been determined to be irreversibly lost and where circulatory and respiratory function has been determined to be irreversibly lost. Furthermore, many jurisdictions that do solely rely on a brain based standard for determining death have allowed it to be inferred as having occurred after a period of continuous loss of circulation under certain circumstances.

The key problem with trying to base determinations of death on loss of circulatory and/or respiratory function is that there is no certainty around the length of time that loss must continuously occur for its retrieval to become impossible. Clinical experience and case studies have already shown that even with patients who are coming off ventilatory support 10 minutes continuous loss is not enough to assure that function will not return (Petrie et al., 2001); indeed there is at least one instance of 20 minutes proving to be insufficient (Maleck et al., 1998). Yet most jurisdictions in which this practice is carried out have national rules that treat loss of function as irreversible and allow related transplant activity after either 10 minutes (e.g. Austria, Czech Republic and Switzerland) or even only 5 minutes (e.g. UK, Belgium, France, the Netherlands, Spain and Canada). Those that leave it to local centres to decide are capable of ending up with some even smaller so-called ‘no touch’ time frames. In the US, for example, local decision-making has resulted in many centres use adopting a 2-minute interval as advocated by the Pittsburgh protocol, and some centres have been known to rely on an even lesser interval on some occasions – 59 seconds in one instance in Denver being the shortest (see Boucek, 2008). These lower time intervals are not even long enough to guarantee that cardio-respiratory function will not spontaneously restart (Price, 1997). Some authors have suggested that loss of function can be considered irreversible when the clinician at hand determines that there is no basis on which to justify reversing it. However, this is merely a desperate sleight of hand contrived to reclassify those who are expected to die within a short period of time as already dead (see further Veatch, 2008).

Even if one is only using a period of loss of circulation to infer brain death, problems emerge. Terminal spreading depression in cerebral grey matter ought to be considered a necessary precondition for a determination of brain death, but there is no clinical or experimental evidence with which to assure us that a specific duration of loss of circulation will have induced such depression (Dreir, 2018). Scalp electroencephalographic (EEG) recordings only vaguely hint at what might be happening to the brain after circulation ceases (Norton et al., 2017). Observations of animals suggest that EEG silence simply indicates the onset of an austerity response to global ischaemia where neuronal energy usage is curbed via the shutdown of nonessential cell functions. Tissue recovery prospects have been shown to endure after this until the neuronal adenosine triphosphate pool depletes (Somjen, 2004).

There appear to be two biological phases that must be passed through after EEG silence before one can start to consider that the brain has actually ceased to function: firstly, a terminal depolarization wave causing failure of membrane pumps *and*, secondly, the resultant cytotoxic processes (Dreir, 2018). The austerity response of the brain to extreme conditions can also be understood as an element of its wider holistic tendency toward preservation and growth (Goldstein, 1939). This tendency in turn is capable of being therapeutically enhanced. It has long been known, for example, that a patient whose heart has stopped beating for 15 minutes after a heart attack can recover if he is treated by cooling the body to 33°C, cardio-pulmonary bypass, cardioplegia (stopping the heart-beat chemically) and a slow increase in oxygenation for 24 hours (Weisfeldt et al., 2002). Up to 80% of these patients can be discharged from hospital, 55% having a good neurological outcome. Self-evidently, there is no valid reason to surmise that patients in this situation are dead.

As I noted at the beginning of this section, patients have interests in being viewed as alive until it can be confidently be said that they are dead. These interests are blatantly being diluted to meet the end of procuring organs for transplantation in this context – the shorter the time waited between loss of circulation and declaring a patient to be dead the less the warm ischemic time and thus the more the potential, all other things being equal, for their organs to be in a good state for transplantation. It is notable that many clinicians will isolate cerebral circulation when restoring the supply of oxygenated blood to retrievable organs after death has been pronounced following the loss of circulation, suggesting that they are not actually confident that the death of the brain can be inferred from such loss. Overall, the situation was already being described at the turn of the millennium as one of thrift euthanasia being conducted with impunity on a national and international scale (Ducharme, 2000). That was and still is a fair characterization when one considers how a range of jurisdictions are allowing medical professionals to treat human beings as dead and remove their essential to life organs in situations where as a matter of precaution they ought to be treated as still being alive at the point of such removal and thus as having been killed by it.

In rounding off this part of the discussion, it is worth noting that in 1974 the famous philosopher Hans Jonas noted that we do not know with certainty the borderline between life and death (Jonas, 1974). Exoterically at least, the same is true today and as he also pointed out a definition cannot substitute for knowledge. For the reasons I have already suggested, I agree with his assertion that in the absence of certainty we should 'lean over backward toward the side of possible life. There is no convincing reason why in theory we should equate death to irreversible loss of brain function or loss of circulatory function (on its own or in combination with loss of respiratory function). Reliance on these equations in practice has been widespread both in spite of this and of the fact that it has proved impossible to be certain as to when either function is irreversibly lost.

Patients who are defined as dead upon such criteria ought to be treated as alive and as entitled to respect for their worth in the same way as other living persons. If they are not in a state to make their own decisions their interests must be protected and given that their health state is likely to be perilous it is difficult to see any, let alone many, circumstances in which their use for transplantation is going to be justified. Whilst this is recognized and accepted in parts of the discourse, jurisdictions which carry out transplantation have routinely gone to the other end of the spectrum of allowing essential to life organs to be removed from such patients in spite of their overt commitment to the Dead Donor Rule. They are reinforced in this by the fact that much of the discourse had professed to be grounded on respect for human life and worth whilst simultaneously defending and in some cases extending the scope of radical approaches to defining death. Furthermore, some of the discourse that does criticize some or all current approaches to defining death does so from a stance of commitment to an ethical position that is incompatible with respect for human worth and life. Some thinkers, for example, recognize that current approaches to death do not properly defend life but argue that individuals should be able to give (advance) consent to even life-ending removals of their organs and/or other materials for transplantation (see, for example, Miller & Truog, 2012). Equally, reflecting personistic and/or utilitarian thought, some of the discourse rates the worth and life of such patients behind the perceived value of removing their materials for transplantation (see, for example, Harris, 2003; Emson, 2003).

Even leaving aside questions about whether transplantation actually has value and what the state of patients who are not actually dead in this situation will actually be, this type of ethics ought to be rejected for reasons already outlined in Chapters 1 and 2. The fact that it is a view which in effect has triumphed for now reveals something disturbing not only about the way medicine is operating in practice but also about its regulation. Many jurisdictions have shown undue deference to prevalent medical approaches to death. In the UK, for example, there has been no effort to legislatively constrain the parameters of how death should be approached and case law on the subject has endorsed the principle of it being verified by practitioners according to standards laid out in professional guidance. The latest example of this is the case of *Re: A (A Child)* [2015] EWHC 443 (Fam) (1) in which the High Court accepted that it could be verified by reference to the 2008 Academy of Medical Royal Colleges Code of Practice for the Diagnosis and Confirmation of Death. Notably, this was in spite of the fact that this Code extended the scope of who could be treated as dead considerably by detailing circumstances in which brain death can be inferred following 5 minutes of continuous loss of circulatory function following withdrawal from ventilatory support. The US Uniform Determination of Death Act is not much better because it does not address what conditions need to be met to satisfy the irreversibility standard contained within section 1 of it.

It is, unfortunately, not just the approaches taken to death that are in hock to transplant need but also the approaches taken to those who have been defined as dead. This is reflected in the fact no jurisdiction with law on this issue confine their use to this end to circumstances in which they have given consent or someone nominated by them to decide has given consent. Rather, all also leave some scope for their use where they did not make a decision or nomination. A tripartite typology can be used to broadly categorize differences in scope provided:

(1) So-called weak (or soft) opt-in systems provide scope subject to a consent being provided by a relevant member or members of kin.
(2) So-called weak (or soft) opt-out systems provide scope if there is no objection by a relevant member or members of kin.
(3) So called hard (or strong) opt-out systems provide general scope – though only for pertinent classes (there may be, for example, exclusions or limits on their application to those who are not long term residents or who had been a minor or long term incapable adult at the point of death being pronounced).

The recent 'mood music' of legal reforms has seemingly been unidirectional towards opt-out – indeed, actually towards various forms of system that may be characterized as hard opt-out. Via the Human Transplantation (Wales) Act 2013, Wales introduced a scheme from 2015 that allows consent to be deemed in relation to the performance of transplant related activities on most adult residents who died prior to making a decision about them/nominating a person to decide for them. The scheme could be characterized as strong opt-out in as much as does not give any relative a binding power of objection. The same is also true of the very similar scheme that will come into effect in England in the Spring of 2020 through changes to the Human Tissue Act 2004 instituted by the Organ Donation (Deemed Consent) Act 2019. Both of these schemes do at least require input from kin to be accounted for in so far as it is relevant to determining what the deceased would have wanted, which is more than can be said of the either the new Colombian law which came into force in 2017 (Law 1805 Desde hoy la donación de órganos es obligatoria) or the reform Bill currently being considered by the Scottish Parliament (Human Tissue (Authorisation) (Scotland) Bill) 2019. However, this is not to say that the Welsh or English schemes could not have gone further to ascertain the wishes of the deceased even within the framework of a hard opt-out approach. They could, for example, have required specific and repeated communication with each and every member of the populace affected in order not just to tell them about the reform but to encourage them to use the register to ensure their specific preferences were still prioritized in practice. This is what the Netherlands is doing prior to bringing in its hard opt-out approach from the summer of 2020 (via its so-called Active Donor Registration Law, 2018).

There are a number of reasons to prefer even a weak opt-in approach to not just a hard opt-out scheme but even a weak one. The most important of these is the simple fact that it at least entails getting agreement or consent from someone. However, the extent to which it allows proxy consent still renders it inconsistent with respect for the worth of the deceased. The normal justification for one individual to make a decision on behalf of another is not merely that the former has the capacity to do so and the latter does not but that the former is also in a position to be able to protect the interests of the latter. Yet the dead have no therapeutic interests, and their interest in being free of unnecessary intrusion could only be outweighed by an interest in donating where it is clear that they had committed themselves to it after having formed a clear understanding of its implications. Since such a form of commitment could only have been demonstrated by them having given consent in the first place, there is no appropriate role for proxy decision-making other than where a power to act as proxy decision-maker was specifically donated by the deceased.

Weak opt-in has also proved more problematic in practice than it is in theory because of the fact that health services in jurisdictions that use it, are tending to treat rudimentary pre-mortem registrations of wishes in favour of post-mortem donation as if they were a proper agreement or consent to such. The English system as I write (i.e. prior to the reforms scheduled for Spring 2020) is a good example. Like many systems, it uses multiple fora to solicit donation. One of these is the online organ donation register available at www.organdonation.nhs.uk (see Figure 5.1). The landing page for this site starts with a big red and big green box alongside each other with respective icons one can click for 'NO, I do not want to donate my organs' and 'YES, I want to donate my organs.' In other words, one is invited to make a choice without first being informed of any of the implications of doing so. If you click 'YES,' you arrive at a page which will result in you being registered after you have filled out some fairly basic personal details and clicked either an icon specifying that you want to donate all of your organs and tissues or one that says some (which then allows you to click on further icons specifying which). There is nothing you are required to do that will ensure that you are filling in the form for yourself rather than for someone else (even potentially unbeknownst to them). Furthermore, whilst there is the option to click on an icon labelled 'More information about your choices,' doing so simply presents you with information about the range of choices you have, not the implications of making those choices. Another method used to recruit donors is the Government online system for driving licence application and renewal (follow applicable link from www.gov.uk/browse/driving/driving-licences). This propagates the strictly inaccurate notion that organ donation saves rather than extends lives, fails to give any information about it whatsoever and when it solicits registration manipulatively refuses to take no for an answer, instead providing only a choice between registering now or being asked again about it at the end of the application at which point one is given another manipulatively structured choice of registering or not registering now. This is in spite of the fact that, since Wales introduced an opt-out law in 2015, NHS Blood and Transplant has had in place the technical capacity to accept refusals.

DVLA is working with the NHS to encourage organ donation and save lives

If you needed an organ transplant would you have one? If so, please help others.

Yes, I want to join the NHS Organ Donor Register, then continue with my DVLA driving licence application. I'd like to complete my DVLA driving licence application first and find out more when I have finished.

If you have registered before, please select 'Yes' and take this opportunity to confirm/update your donation preferences.

To continue with your DVLA licence application, you must select one of the two options above.

From the DVLA website: https://www.gov.uk/browse/driving/driving-licences .

Figure 5.1 Manipulative touting for donors in driving licence applications and renewals

Treating people as having given consent to donating their organs when they have been kept in the dark about the implications of doing so, is inconsistent with respecting their worth and with the normative emphasis that health and human rights law place on informed consent to medical intervention. In some cases, it is also specifically inconsistent with how transplant law treats the process of consent to donation (see Section 5.5). In England, for example, the position as it now stands at least is that consent to cadaveric donation can only be relied on where it is informed consent (Human Tissue Authority Code of Practice concerning consent – Code 1 of July 2014, paras 33–35 – see www.hta.gov.uk). Furthermore, it is also disturbing due to the fact that some of the risks posed by being viewed as a potential donor are less than obvious but at the same time significant. My own ad hoc experience of asking law students taking a medical law module about transplantation is that few have prior knowledge of the fact that both prior to and after death being pronounced a person who has not been excluded as a potential donor can end up being subject to one or more interest affecting interventions or other practices that are aimed at maintaining the potential for their use and maximizing its potential effectiveness (see further below). Equally, few have prior knowledge of the flawed approaches that are currently being taken to the determination of death and consequently of the risk that being viewed as a donor poses of even ending up being killed (by one or more essential to life organs being removed from one's body when one is as a matter of fact still alive).

Overall, I am sceptical about the near term prospects for systems being developed in a way that centres on proper pre-mortem consent. My own prior efforts to make the ethical and human rights case for this (see particularly Garwood-Gowers, 2013; Garwood-Gowers & Pereira, 2017b) seem to have fallen on largely barren ground. Respect compromising attitudes towards human beings in general and the dying and dead in particular have combined with a zealous focus on expanding procurement for transplantation to create a 'perfect storm.'

It is now fairly routine for human beings who are viewed as having the potential to be used for transplant purposes after their death has been pronounced can end up both before and after this point being subject to interventions and other practices to facilitate it that will affect and potentially damage their interests. To help maintain a person's physiological stability and thus organ quality whilst awaiting the outcome of brain death tests, those who are regarded as prospective 'donors' can end up being subject to a range of interventions which in the given context are non-therapeutic and in some cases physically damaging. These include the likes of administration of fluids and inotropes, insertion of central venous and arterial catheters and adjustments to mechanical intervention. The care environment may also be affected. For example, someone admitted to an emergency department with a brain injury that has been diagnosed as grave and incurable may nonetheless be admitted to an intensive care unit and maintained on a ventilator for a few hours in order to preserve the potential to use their organs after death has been deemed to have occurred. In contexts of loss of cardio-respiratory function, human beings that are regarded as prospective organ donors can end up being subject to even more significant alterations in how they are treated in order to facilitate their ultimate use value for transplantation. Under local and/or national protocols, they are likely to be subjected to a range of interventions both prior to and after death has been pronounced that are designed to maintain the quality of their organs. Some of the interventions performed prior have the potential to significantly or in some cases even gravely compromise the interests of those subject to them. They can, for example, extend to the administration of cooling liquids via a catheter and the potentially dangerous administration of blood thinners.

From a worth based ethical perspective, the most fundamental problem with these interventions and practices is that they add grist to the mill by being ways of facilitating the procurement of organs on the back of deeply flawed approaches to death. Secondarily, whilst a worth based perspective and the norms of both healthcare law and human rights coincide in emphasizing the importance of enabling people to make their decisions wherever possible, the design of transplant systems is such that the issue of potential performance of these types of intervention and other practice on a specific individual only tends to be raised when that individual is unable to make or even have input into the decision due to already being either in a coma or 'dead.' Governments, national health services and medical professionals involved with transplantation are quite capable of forthrightly telling the general public and, in particular, those thinking of signing up to donate about the existence of these interventions and other practices and their nature and potential impacts on our interests. They choose not to do so out of fear of putting us off donation and yet at the same time are contradictorily prepared to advocate an expansively transplant friendly approach when it comes to interpreting our interests in the course of making decisions for us in relation to the performance of these interventions and other practices. As I have asserted elsewhere with my Portuguese colleague,

> (t)hese determinations are normally grounded on nothing more than speculation as to what they might have wanted abstracted from whether they had previously expressed a wish in favour of or against donation and, absent having done so, abstracted from the even greater distance of an assessment of their overall character traits.
>
> (Garwood-Gowers & Pereira, 2017, 413b)

The living and procurement

As I observed earlier in this chapter, the first successful kidney transplantation involving a living human donor occurred in 1955 and since then it has become increasingly commonplace – actually further accelerating from the turn of the millennium such that by 2016 it made up 40% (36,149) of total kidney transplants. The other for of living organ donation that will be focused on in this section is liver segment which started to be practiced in the 1980s and has also accelerated its momentum since the turn of the millennium such that in now makes up 20% of liver transplants (WHO/ONT, 2019). Lung lobe, small bowel segment and pancreas islet are other examples of parts of an organ that can be live donated but are not commonly practiced.

When I started looking at living donation in the early 1990s when there was a great deal of resistance to relying significantly on living donors for kidneys because of the harm and risk caused to the donor. Liver segment was in its relative infancy but met even greater resistance because it was becoming clear it entailed even more significant levels of harm and risk. There were also some concerns around the extent to which decisions to agree to these forms of donation could be relied on as proper consents. These concerns have never really gone away. In this section, I join them fully together for the purposes of analysis for the first time. Doing so reveals how mutually aggravating they are and brings into question the reliance now placed on living donor organ transplantation as a whole.

First, on the harm and risk issue, living donors suffer some inevitable harm to the body from the process of undergoing anaesthetic, being cut open and having the relevant material cut out. They also risk perioperative complications (those arising from work up to the surgery, the surgery itself or post-operatively) and suffering ill effects (usually in the longer term) from living without the material that has been extracted. These are 15–20% for living kidney donation. A recent national study in the US (Lentine, 2016) found a reported rate of 16.8% across 14,964 living kidney donors from 97 centres, composed of gastrointestinal (4.4%), bleeding (3.0%), respiratory (2.5%), surgical/anaesthesia-related injuries

(2.4%) and 'other' complications (6.6%). Living with one kidney increases the risks of hypertension, diabetes and end stage renal failure itself. It may also increase the risk of cardiovascular death and of chronic kidney disease (as defined by a low glomerular filtration rate or proteinuria).

Perioperative complication rates for living liver segment donation are at least double those for living kidney donation. A recent US-wide study (Abercassis, 2012) put them at 40%. The study examined 740 people who had donated on average 3.4 years earlier (a further 20 went into the operating theatre with the intention to donate but were excluded because of health issues that became apparent). Of these, 296 suffered one or more complications. There were 557 complications experienced as a whole with the most commonly occurring being infections (112), pleural effusion (83), bile leak/biloma (62) incisional hernia (54), psychological difficulties (51), ileus (25), neuropraxia (24), ascites (21), unplanned reexploration (20) pulmonary edema (15) and bowel obstruction (13). Perhaps even more significant is that whilst the reported incidence of serious complications in kidney donation is little more than a few percent (Mjøen, 2009; Lentine, 2012) it seems to be around 15–20% for liver segment (Kousoulas, 2011; Abercassis, 2012). Since its creation in 2004, the Clavien-Dindo Classification has increasingly been used to rank surgical complications on a scale of 1–5 according to their level of seriousness (Clavien, 2009), with 1 being essentially more minor complications and 2 upwards being serious with 5 being death. The aforementioned national study in the US found that nearly half of the complications suffered (277 out of 557) reached level 2 and upwards which meant that they were at least potentially life-threatening (Abercassis, 2012). Notably, liver segment also generates more risk of death than kidney donation – probably by a multiple of 5 or 10 times (around 0.15–0.3% versus around 0.03%). Long term impacts from loss of liver segment are not known.

On a scale of how relevant these serious levels of harm and risk are treated as being, libertarians and those who believe in the above all else do no harm (*primum non nocere*) principle occupy virtually polar opposite ends of the spectrum. A libertarian position will at most give credence to the idea that it is relevant to children and adults who are unable to make their own decision. However, I have already made the case for rejecting this position in favour of respect for worth. An above all else do no harm principle involves taking a conservative approach to harm and risk, ensuring – amongst other things – that medical intervention is only performed where the harm and risk it entails is being contained as far as is possible and is clearly outweighed by the prospective benefits that flow from the intervention.

The main tussle over the nature of the principle concerns whether that benefit must take a clinical form that directly stems from the intervention or whether it could also be non-clinical and/or indirect. Certainly some clinicians continue to regard it as excluding the latter, and in surveys, a very small percentage who are asked about the acceptability of living donation oppose it on this basis (Pascalev, 2013, 32). However, others regard the principle as capable of accounting for all potential benefits to the prospective donor. This opens up the capacity for living donation to occur when it provides a net benefit to the donor. Perhaps the most noted advocate of this position in the context of organs is Spital (see, for example, Spital, 2006; Spital & Jacobs, 2007, 2008). He adds the proviso that the donation should be unlikely to cause grave harm or death, but this may be considered almost redundant since it would hard to picture any form of living donation being of net benefit were it not to meet this criteria.

Both of these perspectives tend to be drawn as much if not more from specific notions of professional ethics as they are broader conceptions of ethics in general – in other words, on the notion of medical professionals being advocates for the interests of their patients or, to be more precise in the prospective donor context, volunteers. There is also a potential for healthcare to be adversely affected if medical professionals deviate from that role – not least because the public will have reduced confidence in them. Two further points made persuasively in the donation context by such authors as Elliot (1995) are that just because a person wants to do something does not per se mean that another is per se morally justified in helping them to accomplish it and just because someone wants to give something doesn't make it morally justified to accept it.

These points are significant because they provide a cogent basis on which to reject a libertarian stance to donation (Williams, 2018). However, they do not specifically mean that an approach based on the above all else do no harm principle should be adopted. Even the net benefit to the donor version of such a standard would be very paternalistic – leaving those competent/capable to decide to donate with little more leeway to do so than those who are not. Spital's counter-argument that it is valuable to bring standardization to the treatment of the different classes (Spital, 2004) is weak. By contrast, arguments for prohibiting medical use that hits the threshold of being exploitative are strong. I would argue that extraction of material from the living for transplantation hits this threshold not just where it entails harm and risk that hits any of the three markers of being:

(1) Unnecessary,
(2) Excessive per se or
(3) Not exceeded by prospective benefit to the donor and/or others.

Broadly speaking, between them governance and law relevant to living donation tend to embody these requirements. The 2010 WHO Guiding Principles on Human Cell, Tissue and Organ Transplantation, provide some steer by emphasizing (in Guiding Principle 3) that donation from deceased persons should be developed to its maximum therapeutic potential and that when living donation is considered, it is under selection criteria that are scrupulously applied and monitored and that when it takes places donors are provided with professional care and well-organized follow-up. However, much more detailed leadership on this issue has come from the Council of Europe via the Additional Protocol to the CHRB Concerning transplantation of organs and tissues of human origin (AP).

Art. 9 of this stipulates that tissue and organ removal for living donation should only occur when there is no suitable material available from a deceased person *and* no other alternative therapeutic method of comparable effectiveness. Para. 59 of the Explanatory Report to the AP contextualizes this requirement by noting that 'removing organs or tissues from living donors for implantation purposes always has consequences and may carry some risk for that donor' adding that this implies that it should not be done 'where an appropriate organ or tissue from a deceased person is available.' Common sense suggests that what is being referred to here is not necessarily immediate availability but rather likely to become available within a period of time that is reasonable given the health exigencies of the situation. Thus, one might engage in a wait for cadaveric where the chances of obtaining it quickly enough to avoid material damage to health were good, but in other situations generally and emergencies in particular, one might reasonably proceed to living donation provided that the second cumulative requirement was met. That second requirement requires an openness to alternatives to living donation. However, para. 60 of the Explanatory Report specifically makes it clearer that dialysis is not considered to be of comparable effectiveness, and para. 61 suggests that cadaveric would not necessarily be either where living produces superior results. Cure is not mentioned in either paragraph but ought to be reflected on and has already proved itself pertinent in the context of renal problems brought on by Type 2 diabetes – viz the potential explored earlier in the chapter to ameliorate and even reverse it via changes to diet and exercise regimes.

Art. 11 requires prospective donor to be properly physically and psychologically screened and then protected from use where their health or any other factor renders them unsuitable. It also embodies an emphasis on containing overall levels of harm and risk, by stating that '(t)he removal may not be carried out if there is a serious risk to the life or health of the donor.' The nature of what constitutes a serious risk is not specified, but it could reasonably be defined as a risk that was serious either because it was a significant risk of a significant adverse outcome or a low risk of a very serious adverse outcome.

At the domestic level, similar constraints often exist by direct virtue of specific transplant legislation. For example, Lopp's work has shown that in Europe:

- Many pieces of such legislation require the consideration of other treatment options for the prospective recipient, and some of them go on to require living donation specifically to be the only suitable option or the most efficacious or efficient (Lopp, 2013a, 129).
- Many place limits on harm and/or risk directly and/or via restricting the types of material that can be donated (Lopp, 2013b, 31–66).
- Most also require the all-round prospective benefit of donation to outweigh the harm and risk it generates (see further Lopp, 2013b, 31–66).
- Some place restrictions on who solid organs can be donated to in an effort not just to lessen the risk that donation will be commercially motivated and/or involuntary but also to strengthen prospective donor benefit (Lopp, 2014, 161).

There is no up-to-date survey of what transplant laws do in other parts of the world, but certainly older surveys suggest that they at least tended to take a broadly similar approach (see, particularly, Garwood-Gowers, 1999, 50–61). Furthermore, all jurisdictions will have one or more overarching legal principles that can be used, amongst other things, to contain what one person can do to another, notwithstanding that other might be a capable and willing 'victim.' For example, Germany has a recognition of dignity in Article 1 of its Basic Law and also a more specific provision in section 228 of its Criminal Code that a person who injures another cannot use consent to defend themselves from criminal liability where there action has been contrary to good morals. Under the common law, a purported consent to a medical intervention that is considered unreasonable or improper will be deemed invalid on public policy grounds. Whether such provisions will be treated as violated by living donation that falls short of one or more of the three aforementioned markers is moot as are the precise implications for living donation if they are. There some common law authority to suggest that removal of a vital to life organ in a non-domino situation would be illegal (*obiter dicta* in the Canadian case of *Urbanski v Patel* [1978] 84 DLR (3d) 650), but we do not know whether the Courts in common law jurisdictions and elsewhere might be more astringent than this and start questioning some of the forms of living donor organ transplantation that do not involve removal of a vital to life organ but nonetheless impose serious harm and risk. An obvious candidate for such astringency would be lung lobe donation – especially in view of the fact that it entails two donors. On the other hand it is hard to see a Court impugning removal of a kidney or liver segment for donation unless it starts to take on more fundamental criticisms of transplantation or the situation was irregular in a problematic fashion, such as the profile of harm and risk to the donor being unusually high and/or the expected benefits to the prospective recipient being unusually low. Nonetheless, even leaving fundamental criticisms of transplantation aside, we should still be concerned with even run of the mill forms of kidney and liver segment donation because of the fact that the still significant levels of harm and risk they pose are rendered more problematic by the fact that decisions to donate them are typically poorly made – a problem to which I now turn.

In considering the quality of donor decision-making, it is important to start by reflecting on what constitutes a good decision-making process in general. Traditionally rational models of decision-making have been viewed as an ideal. These entail the decision being broken down into a series of steps that start with surveying the full range of objectives to be fulfilled and canvassing alternative courses of action, continue with evaluating the pros and cons of each and end up with making a final choice and plans for its implementation or execution (Janis & Mann, 1977). However, increasingly empirical evidence is pointing toward an intense direct focus on information actually being counterproductive to good decision-making. For example, one key study found that participants who performed a distracter task before making a decision generally performed better at making it than those who were asked to get straight into thinking very carefully about the information they had been given (Dijksterhuis, 2004). This and other recent studies have posited that a more diffuse state gives room for processing to occur beneath the level of consciousness – fostering more advanced use of information, including better polarization of the alternatives and better clustering of the information relevant to each of them. One may hypothesize that this, in turn, provides a better platform for the distillation of focus from relevant information as a whole to the most critical pieces of information in particular – something that could be posited to be significant to good decision-making given that experts usually ultimately base their judgments on a few pieces of information (Shanteau, 1992).

However, donors cannot be characterized as engaging in good decision-making processes even on this opened-up view of what that entails. Of 98 kidney-to-kin donors in a 2005 Vancouver based study, 87.7% decided almost instantly to donate when they found a kidney was required (Stothers et al., 2005). Earlier empirical studies produced similarly troubling results (Fellner & Marshall, 1968; Simmons et al., 1977; Garwood-Gowers, 1999; Lennerling et al., 2003) as have empirical studies of liver segment donation (for a systematic review see Gordon et al., 2011). The majority of those who donate either to kin seem to do so prior to properly knowing, let alone understanding, the implications of doing so. It is likely that this is partly attributable to the way the social description of organ transplantation interacts with the psychology of many prospective donors.

I have already described how little attention is given to whether organ transplantation is acceptable and productive as a therapy or to prevention and cure of the diseases that feed into the perceived need for it. Instead, it is simply treated as the gold standard treatment for certain conditions and the only effective treatment for many others and often imprecisely propagated as being life-saving rather than life extending. Where gaps between demand and supply of it exist, it tends to be reflexively treated as in need of greater supply to avert adverse health consequences. At the individual level, wherever there is a prospect of a prospective recipient of a kidney or liver deteriorating or even dying whilst waiting for a cadaveric organ to turn up, there is almost an expectation that living donation will be considered. Furthermore, there is almost a specific expectation that where those asked to donate are kin to the prospective recipient that they will offer immediate psychological and practical relief to his or her 'desperate plight.' Indeed, prospective recipients may be naturally inclined to view reflection by those asked as suspect – a marker of ambivalence to the idea of donation and their well-being as a whole. The following extract of an interview I conducted as part of the EUROTOLD project with a Norwegian academic who had received a kidney from her father, highlights the bind donors face in this respect. She wondered whether his offer was down to his 'terrible sense of duty' but then proceeded to reassure herself that it probably was not by recalling the reactive immediacy of it:

> I knew that things could go wrong [for her father from the surgery] and I thought that would be absolutely awful – I couldn't live with that but then I sort of remembered the situation at the summerhouse. I was sitting on the terrace over breakfast and I sort of put this question sort of timidly

> and I just remembered they both jumped in their chairs and said 'please I want to do this.' This spontaneous reaction was my proof that this was really something they wanted to do and not something they felt they should do from any kind of negative duty, so I sort of cling to that (laughing). That was very important to me.
>
> (see Garwood-Gowers, 1999, 97)

Logically, one would think that the impulse to decide immediately would be present almost whatever the main motivations were for donation. It seems that a significant minority of people want to donate at least partly to make up for what they consider to be wrongs done in their life (around one-quarter of 114 donors in the Minnesota study – Simmons et al., 1977, 162) and that whilst this may mean that donation is partly about redemption in a general spiritual sense it is also often likely to be about squaring things up with and/or appeasing family (29% of donors in the Minnesota study agreed that they 'had done something major in their life that their family did not approve of' – Simmons et al., 1977, 162). It is also likely that the desire not to lose face with family and to avoid feeling guilty or otherwise self-critical are factors.

A desire to fulfil one's duty can also be a factor as evident in the interview with the Norwegian academic and her father. In this context, the motive to donate can bear resemblance to what the social science literature calls rescue altruism. In the context of aquatic rescue Pearn and Franklin (2012) have summarized this phenomenon as being composed of four elements: (1) an ethos based on the Good Samaritan or Golden Rule ethic; (2) a subjective identity of the rescuer with the victim, intensified by a perceived duty-of-care relationship; (3) perception of risk in which the potential of successful rescue is greater than zero; and (3) personal courage that ignores degree of risk. Their study of 103 people who drowned in Australia over an 18-year period whilst trying to rescue a related (67%) or unrelated (33%) person or persons from water indicates how far people are prepared to put their physical well-being and even life at risk in the course of rescue – none of those who died were professionally trained in aquatic rescue, and only half of them were familiar with the water hazard. Whether many people are prepared to put their life in serious jeopardy for the cause of organ transplantation is moot, but we can surmise from living liver segment and kidney donation statistics that many are prepared to expose themselves to considerable harm and risk in this context.

The fact of decisions to donate being impulsive is rendered more problematic by the fact that many will find it psychologically difficult to feel free to withdraw their agreement, especially in the context of donation to kin. Specifically, the sense of guilt, of having falsely got everyone's hopes up and wasted their time, and of being left as the 'black sheep' are all likely to work as significant impairing factors in many cases. Some empirical evidence suggests that the longer the process progresses, the more difficult it is to withdraw (Simmons et al., 1977, 260). There is a clear danger that those who have decided to donate will 'push down' any feelings and thoughts which open that decision for question and find ways to reify having made it even to the point of constructing a falsely rosy picture of it. Within the interviews and questionnaires I and others conducted for the EUROTOLD project, there was significant evidence of such reification taking place, most powerfully the fact that 19 of 26 donors responding to a question asking for quantification of risks answering that it had none (Garwood-Gowers, 1999, 67). The slightly more recent Vancouver study also adds to a picture of potential reification with 74.2% of donors saying they were not worried about the long term effects of living with one kidney (Stothers et al., 2005).

It also seems to be the case that weaknesses in the decision-making process are not being properly mitigated and in some cases are being exploited and exacerbated by the character of input that comes from the medical side. As a context to discussing this, it is useful to note that medical organizations where donation takes place and medical professionals who carry it out have a duty to foster good decision-making with respect to it. One element of this is that they will be subject to legal and broader governance based duties to ensure that those who are considering donation are equipped with appropriate information pertaining to such matters as its risks, benefits and consequences and the alternative options to it.

The precise nature of these duties will vary (particularly) across jurisdictions, but many legal standards are organized around the need to give decision-makers all the information about the nature, purpose, risks and consequences of the proposed intervention and relevant alternatives to it that they would reasonably need in the given situation to protect their interests. That information is likely to be very extensive in the case of non-therapeutic interventions particularly when – as is the case with living donation of kidney and even more especially liver segment or lung lobe – they cause significant harm and impose significant risk on those who undertake them. The extent to which this duty to disclose is being fulfilled is open to question.

The largest survey of donor experience and attitudes that has ever taken place – namely 484 donors in Norway (Westlie et al., 1993) – presents a picture of the majority being unequivocally satisfied with how much information they were provided on how they were selected as a donor, what surgery would be like and what the chances of success for the recipient were. Their opinions were only much more mixed when it came to how much information they were provided about what post-surgery hospitalization and recovery at home would be like. However, part of the whole point of disclosure to a person making a decision about medical information is that they are generally largely in the dark about its implications. Data as to whether they satisfied with the extent of disclosure that has taken place is thus an inherently unreliable indicator of whether or not the disclosure that took place was actually appropriately extensive. A better approach is to assess what clinicians themselves what they disclose. One study that has recently done this with respect to risks in kidney donation involved 32 surgeons from across 8 centres in the Netherlands (Kortram et al., 2016). Table 5.1 illustrates the percentages which were achieved in terms of frequency of disclosure with respect to risk relating to mode of procedure, short term and long term complications and living with one kidney:

It is evident from this that there were not only significant variations between surgeons as to what they disclose but also significant overall weaknesses in overall levels of disclosure. Half of the surgeons, for example, only routinely disclosed the small but very serious outcome of death and disclosure rates were even less

Table 5.1 Results on risk disclosure

Always	*Always*	*Sometimes*	*Never*
Conversion to open procedure	84.5	6	9.5[a]
	Short term complications		
Bleeding	100	–	–
Wound infection	97	3	–
Pain	84	6	9
Fatigue	53	22	25
Damage to other organs	50	31	19
Pneumonia	44	28	28
Abscess	31	28	41
Urinary tract infection	31	34	34
Thromboembolic complications	28	34	38
Testicular pain/swelling	25	22	53
Cardiovascular complications	15.5	44	40.5
Neurapraxia	12.5	40.5	47
Neuropathy	6	47	47
Mortality	50	37.5	12.5
	Long term complications		
Incisional hernia	59	28	13
Kidney failure	47	28	25
Chronic pain	28	44	28
Hypertension and other cardiovascular complications	13	34	53
	Risks of living with one kidney		
Smoking	37.5	22	40.5
Obesity	19	47	34
NSAIDs	15.5	37.5	47
Antibiotics	13	22	66

Source: From the Kortram et al., 2016, study.

for several other quite serious short term risks and also most of the key long term risks and all of the key factors exacerbating the risk of living with one kidney. The overall picture seems to be one of patients' right to know not being taken very seriously both with respect to the making of an informed choice about donation and having the appropriate knowledge to be equipped to minimize risk should they do so (e.g. by maintaining a good weight and avoiding smoking). Given that the Netherlands is probably at least the equal of most countries when it comes to respecting patient autonomy, there must be concerns that the results would be no better were the same study to be repeated elsewhere. There are also questions of whether information is being properly tailored to take into special risk factors where these are present, such as with donors whose health is marginal or whose ethnic background might make them more vulnerable than average to experiencing ill effects from the loss of material. An example is the fact that being an African-American donor carries three times the risk of end stage renal failure as being a Caucasian-American one (Henderson & Gross, 2017).

In addition to this, there are concerns about the extent to which medical professionals are properly discharging what one might describe as their duty to make reasonable efforts to ascertain that what has been disclosed has been properly understood. The EUROTOLD cohort of donor interviews and questionnaires raised grave concerns on this. 31 of 48 donors answered 'none' to a question about the risks they were aware of at the time of donation at the time of donating, and 19 of 26 answered 'none' to a question about the level of risk (see Garwood-Gowers, 1999, 67). Whilst these interviews and questionnaires were executed months and in some cases years after donation took place, the fact that such a large proportion of donors were so clueless about risk cannot reasonably be attributed to problems of recall and rather more suggests serious problems with both levels of disclosure and perhaps also comprehension – the former at least fitting with the context of the fact that some of the interviews were conducted in Slovenia which had only recently broke free of longstanding totalitarian rule. However, this is not to suggest that there are no concerns in cultures with a more longstanding tradition of valuing patient rights. The Vancouver study, for example, found that 74.5% of donors disagreed with the view that immediate post-surgical complications in donation are common and may be life-threatening (Stothers et al., 2005). Disclosure and checking that what has been disclosed is understood may be at such weak levels overall as to render donations short of legal as well as ethical standards, particularly in the European context where the ECHR acts as a backstop for ensuring that domestic standards are robust (see further Garwood-Gowers, 2017, 260–267).

The final issue that needs discussing is how the specific medical set which prospective donors encounter biases them towards donating. There are specifically several different ways in which it influences them toward what is known in the psychological literature as compliance – a phenomenon which Breckler et al. (2006, 307) have described as involving individuals acting in a certain way because another person or group wants them to even though they technically have the option to refuse to do so. Whilst that want may often be overt and transparent when it comes to the prospective recipient it tends to be more subtle and elaborate when it comes to the medical set-up. One element of it relates to the fact that at times people are not always formally asked to make a decision about whether or not they want to donate before they are engaged in the process that leads towards doing so. Deliberately or unwittingly this makes unfair use of the so-called one foot in the door technique – namely the fact often exploited by salesmen and women that people are more likely to perform a target behaviour having performed preliminary behaviours towards it (Freedman & Fraser, 1966; Bandura, 1969).

Not only those who agree to donate, but also those who undertake tests with a view to doing so, tend to find it more difficult to withdraw their involvement once they have done such things as have discussions about it, gone for a blood test and had their tissue compatibility and clinical suitability checked. The medical set-up also at least unwittingly exploits the fact that people can be biased toward 'complying' with a target behaviour by being exposed to influences in favour of it, particularly where these are experts (French & Raven, 1968) and especially where there is uniformity of opinion in favour of it (Asch, 1951) and room for modelling (Bandura, 1969) and social comparison (Festinger, 1954).

All of these factors tend to be powerfully at work in the donation decision-making context (Garwood-Gowers, 1999, 99–102) as does testing which can add credibility to expert views and help obfuscate ambivalence (Bem, 1974). They are likely to feed off the fact that donors have to trust doctors (Stothers et al., 2005) and place significant informational reliance on them. Just how far that informational reliance extends in the donation context in an era of mass availability of information via the internet is uncertain. However, it is interesting to note that 90.3% of donors in the Norwegian study responded that they would go to nephrologists at the recipient's hospital if they needed information on kidney donation and transplantation (Westlie et al., 1993), and whilst one would expect this figure to be much lower today, it is likely that prospective donors will still heavily rely on them and other likely pro-donation sources of information.

To conclude this section, I suggest that living donation of kidneys and particularly liver segment involve levels of harm and risk that are questionable even if taken in isolation, but are certainly deeply problematic when one accounts for the doubtful value of organ transplantation as a therapy and for the fact that decisions to donate (to kin at least) are often made impulsively without proper prior knowledge, understanding and reflection upon their implications and within a context of social and psychological pressures to comply with it as a 'target behaviour.'

Commodification

I use the term 'commodification' in this section in its narrow sense to describe money or money's worth being derived from the use of the body as such. In line with the analysis towards the end of Chapter 1, I treat this as being inherently ethically objectionable on the basis of its being incompatible with human worth, specifically the norm of protecting against instrumentalization. It is a phenomenon which in the transplant context is often described as commercial transplantation and extensively overlaps with trafficking in human beings for their body materials, trafficking in body materials themselves and transplant tourism. These activities are opposed in a range of ways at domestic, regional and global levels (Garwood-Gowers & Pereira, 2017, 396–401). At the global level, one source of opposition is the Council of Europe. As was noted in Chapter 3, Art. 21 of the CHRB opposes financial gain being made, as such, from the body or its parts and is virtually mirrored in the immediate context by Art. 21 of the Additional Protocol concerning transplantation of organs and tissues of human origin, though this also helpfully opposes comparative advantage.

Art. 22 of the Additional Protocol adds a stipulation that trafficking should also be prohibited. Recently this has been built on by the development of the

Convention against Trafficking in Human Organs which came into force in March 2018 and as of 11 March had already been ratified by 7 states and signed by 17 others. WHO has also played a significant role. The fifth of its 2010 Guiding Principles on Human Cell, Tissue and Organ Transplantation stipulates that '(c)ells, tissues and organs should only be donated freely, without any monetary payment or other reward of monetary value' and that '(p)urchasing, or offering to purchase, cells, tissues or organs for transplantation, or their sale by living persons or by the next of kin for deceased persons, should be banned.' The sixth opposes related activities including adverts that offer or seek payment to individuals/their next of kin for body materials and brokering that involves payment to such individuals or third parties.

The two principles dovetail with the WHA 2004 Resolution on Human Organ and Tissue Transplantation (WHA57.18, www.who.int/) which urges all states to take measures against trafficking and the related problem of transplant tourism 'to protect the poorest and vulnerable groups.' Transplant organizations have also been active. Notably, the Transplantation Society and the International Society of Nephrology drove the development of what became the 2008 Declaration of Istanbul on Organ Trafficking and Transplant Tourism. The Declaration not only denounces trafficking and tourism as violations of equity, justice and human dignity (principle 6) but advances a set of positive proposals for how organ transplantation should be organized. It has been accredited with having several positive impacts, including stimulating the strengthening of several domestic laws (Danovitch, 2014). As of 8 April 2019, 122 transplant organizations are signed up to it (www.declarationofistanbul.org/index.php).

Despite such opposition to it, the illegal trade in body materials for transplantation and its related activities have persisted. By 2007, at least 5% and potentially up to 10% of all solid organs transplanted globally were purchased illegally, according to extrapolation from meta-analysis of empirical studies of illegal purchase (Shimazono, 2007). By 2012, Luc Noel, a World Health Organization expert, had more instinctively estimated that the latter figure may be closer to being correct (Campbell & Davison, 2012) which is a reasonable assumption given that a great deal of black market activity can tend to be hidden from view. This translates to over 13,500 solid organs if one abstract from the latest (2016) global figures for solid organ transplantation (www.transplant-obervatory.org).

In part, this may be a reflection of moral ambivalence around preventing commercial dealings. Certainly such ambivalence has contributed to a situation in which states have been selective in their approach to legally and/or practically opposing commodification. Iran has honestly advertised this selectivity – it has had an institutionalized system of commercial living donation of kidneys since the 1990s consisting of the donor receiving a mix of state payment and a period of free healthcare as well as being able to receive money from the recipient. Less honestly, some other countries have adopted a policy of asserting that they are opposed to commodification whilst at the same offering money or money's worth incentives for some types of donation. For example, Israel opposes commodification via section 2 of its Organ Transplantation Law 5768–2008, but section 23b of the law provides living donors with exemption of entrance charges to national parks and nature reserves and for cadaveric donors allows the financing of burial expenses. Not dissimilarly, Spain has committed to donation being an altruistic act via its Real Decreto 2070–1999 but for cadaveric donation has allowed its regional health authorities to pay the basic funeral expenses – and costs of repatriation of the body where the deceased is foreign – in cases where no relevant insurance cover exists.

In addition to this it has become the norm in some countries not to defray or reimburse actual expenses for the provision of some materials for transplantation, such as blood, but to instead provide a fixed payment which in some cases is clearly designed and/or will have the effect of incentivizing donation. Finally, there are instances in which the reach of prohibitions has been deliberately kept short of materials regarded as somehow more minimal in their significance. The CHRB has encouraged this because, whilst Art. 21 of it appears to prohibit the body ever being the subject, as such, of financial gain, para. 133 of the Explanatory Report Para makes it clear that this prohibition is not intended to apply to discards such as hair and nails. Para. 134 states that the buying and selling of these is not an affront to human dignity, but this runs counter to the basic premise of Art. 21 which is that making financial gain from the body as such should be prohibited on the basis of its incompatibility with dignity.

The problem of commercial dealing could certainly be ameliorated to at least some extent by the opposition to it being more consistent and robust. This is not just about changing provisions such as the aforementioned which allow for it in some form or other but also about strengthening international co-operation and building on the Convention against Trafficking in Human Organs with comprehensive binding international instruments to harmonize regulations and improve cross-national regulation (López-Fraga et al., 2014). Furthermore, greater attention must be placed on addressing the causes of the trade, including by drastically strengthening conditions of economic justice and financial security and addressing the alternatives to transplantation such that it would, at the very least, be less of a scarcity when resorted to.

5.5. Conclusions

In the course of this chapter, I have illustrated excesses in the historical and contemporary practice of transplantation and also argued that is problematic in a more fundamental sense because of the way in which it interferes with human nature and form and reflects and reinforces a clinically, socially and culturally counterproductive approach to dealing with the manifested health results of people's issues. What is needed is simultaneously a greater acceptance of those issues and a greater emphasis on addressing the cause of those issues by engaging holistically with the potential for prevention and cure. Forays into prevention, such as in the context of Type 2 diabetes, hint of the revolutionary impact that a prevention first approach to health could have both generally and specifically in the context of reducing some of the kinds of problems that are currently treated as indicative of a need for transplant.

This specific point should be noted even by those who reject fundamental critiques of transplantation since it could form a more general strategy to lessening reliance on it as a technology both in order to save cost and reduce pressure on procurement. In solid organ transplantation specifically, or at least particularly, we also run into another problem, however, that even the supporter of transplantation should account for, namely that both the modes of procurement – from the living and from those defined as dead are deeply problematic. Solid organs extracted from the living for transplantation almost entirely come from those who are legally empowered to make their own decision about whether or not to proceed with it. When it comes to their kin being 'in need,' people tend to decide to donate prior to being informed about the implications of doing so. They are generally not given anything like all the information that they are entitled to, often have a limited comprehension of the information that they are given and are rendered vulnerable to making and holding to decisions to donate that may not reflect their 'true' will, not least because of the very pressured and desperate nature of the situation and the way in which the medical set-up draws them in. The situation with respect to removing solid organs from those defined as dead is arguably worse still because current medico-legal approaches to death are not conservative enough to confidently ensure that those who are treated as if they dead, actually are. As has also been shown, there is also not yet enough singularity of commitment to tackle the problem of commodification in procurement. Overall, solid organ transplantation is a deeply questionable phenomenon, practiced to the extreme and founded on an almost unscrupulous approach to procurement.

References

Abercassis, M. (2012) Complications of living donor hepatic lobectomy – A comprehensive report. *American Journal of Transplantation.* Available at https://doi.org/10.1111/j.1600-6143.2011.03972.x.

Accad, M. (2015) Of wholes and parts: A Thomistic refutation of 'brain death'. *Linacre Quarterly.* Available at https://doi.org/10.1179/2050854915Y.0000000004. Accessed 8.3.2019.

Ad Hoc Committee of the Harvard Medical School. (1968) A definition of irreversible coma: Report of the ad hoc committee of the Harvard Medical School to examine the definition of brain death. *Journal of American Medical Association.* 205, 337–340.

Aminoff, M. (2017) The life and legacy of Brown-Séquard. *Brain.* 140(5), 1525–1532.

Asch, S. (1951) Effects of group pressure on the modification and distortion of judgments. In Guetzkow, H. (ed.) *Groups, Leadership and Men.* Pittsburgh, PA: Carnegie Press, pp. 177–190.

Ballantyne, C., et al. (1989) Effects of Cyclosporine therapy on plasma lipoprotein levels. *Journal of American Medical Association*. 262(7), 53–56.

Bandura, A. (1969) *Principles of Behaviour Modification*. New York, NY: Holt, Reinhart & Winston.

Baumgart, H. (1994) Zur frage der organtransplantation. rundbrief fwr die mitarbeiter der medizinischen sektion am goetheanum. *Aller Welt Michaeli*. 7, 251–254.

BBC Newsnight. (2019) UK teen dies after stem cell windpipe transplantation. 26 February. Available at www.bbc.co.uk/programmes/p0721hnt. Accessed 28.2.2019.

Belkin, G. S. (2003) Brain death and the historical understanding of bioethics. *Journal of History of Medicine and Allied Sciences*. 58, 325–361.

Bem, D. (1974) Self perception: An alternative interpretation of cognitive dissonance phenomena. *Psychological Review*. 81, 506–520.

Bernat, J. (2002) The biophilosophical basis of whole-brain death. *Social, Philosophy and Policy*. 19(2), 324–342.

Bhandari, M. & Tewari, A. (1997) Is transplantation only 100 years old? *British Journal of Urology*. 79(4), 495–498.

Bos, M. (2017) The role of ethics in the early history of transplantation. In Massey, E., et al. (eds.) *Ethical, Legal and Psychosocial Aspects of Transplantation*. Lengerich: Pabst, pp. 15–21.

Boseley, S. (2018) GPs to prescribe very low calorie diets in hope of reversing diabetes. *The Guardian*, 30 November.

Boucek, M., et al. (2008) Pediatric heart transplantation after declaration of cardiocirculatory death. *New England Journal of Medicine*. 359(7), 709–714.

Breckler, S., et al. (2006) *Social Psychology Alive*. Belmont, CA: Thomson Wadsworth.

Brightbill, F., et al. (2009) *Corneal Surgery: Theory, Technique and Tissue* (4th Edition). Amsterdam: Elsevier.

Butler, D. (1998) Last chance to stop and think on risks of xenotransplants. *Nature*. 391, 320.

Campbell, D. & Davison, N. (2012) Illegal kidney trade booms as new organ is 'sold every hour'. *The Guardian*, Sunday 27 May 2012. Available at https://www.theguardian.com/world/2012/may/27/kidney-trade-illegal-operations-who. Accessed 25.5.19.

Champaneria, M., et al. (2014) Sushruta: Father of plastic surgery. *Annals of Plastic Surgery*. 73(1), 2–7.

Clavien, P., et al. (2009) The Clavien-Dindo classification of surgical complications: Five year experience. *Annals of Surgery*. 250(2), 87–96. Available at doi: 10.1097/SLA.0b013e3181b13ca2.

Cooper, D. (2014) Progress in pig-to-non-human primate transplantation models (1998–2013): A comprehensive review of the literature. *Xenotrasplantation*. 21(5), 397–419.

Crawford, A., et al. (2013) A brief history of Corneal Transplantation. *Oman Journal of Opthalmology*. 6(1), 12–17.

Cuo, J., et al. (2006) Time dependent validity in the diagnosis of brain death using transcranial *Doppler: Journal of Neurology, Neurosurgery, and Psychiatry*. 77, 646–649.

Danovitch, G. (2014) The declaration of Istanbul on organ trafficking and transplant tourism: The first five years. In Weimar, W., et al. (eds.) *Organ Transplantation: Ethical, Legal and Psychological Aspects*. Lengerich: Pabst, pp. 24–29.

Dauber, M. (2017) Animal testing and ethics in human head transplantation. *AJOB Neuroscience.* 8(4), 212–214.

Davies, R., et al. (2003) The evaluation of disease prevention and treatment using simulation models. *European Journal of Operational Research.* 150, 53–66.

Decker, O. (2014) *Commodified Bodies.* London: Taylor & Francis.

Deschamps, J-V., et al. (2005) The history of xenotransplantation. *Xenotransplantation.* 12(2), 91–109.

Dewhurst, J. (1988) Cosmas and Damian, patron saints of doctors. *Lancet.* 2, 1479–1480.

Dijksterhuis, A. (2004) Think different: The role of unconscious thought in preference development and decision-making. *Journal of Personality and Social Psychology.* 87(5), 586–598.

DiRECT. (2018) Available at www.ncl.ac.uk/magres/research/diabetes/reversal/#publicinformation. Accessed 8.3.2019.

Dreir, J., et al. (2018) Terminal spreading depolarization and electrical silence in death of human cerebral cortex. *Annals of Neurology.* 83(2), 295–310.

Ducharme, H. (2000) Thrift-euthanasia, in theory and in practice: A critique of non-heart-beating organ harvesting. In Freeman, M. & Lewis, D. (eds.) *Law and Medicine: Legal Issues 3.* Oxford: Oxford University Press, pp. 493–525.

Eggers, P. (2011) Has the incidence of end-stage renal disease in the USA and other countries stabilized? *Current Opinion in Nephrology Hypertension.* 20(3), 241–245.

Elliott, C. (1995) Doing harm: Living organ donors, clinical research and the tenth man. *Journal of Medical Ethics.* 21(2), 91–96.

Emanuel, L. (1995) Reexamining death: The asymptotic model and a bounded zone definition. *Hastings Center Report.* July–August, 25(4).

Emson, H. (2003) Is it immoral to require consent for cadaver organ donation? *Journal of Medical Ethics.* 29, 125–127.

Evans, D. & Hill, D. (1989) The brain stems of organ donors are not dead. *Catholic Medical Quarterly.* August, 113.

Fellner, C. & Marshall, J. (1968) Twelve kidney donors. *Journal of American Medical Association.* 206, 2703–2707.

Festinger, L. (1954) A theory of social comparison progress. *Human Relations.* 2, 117–140.

Fox, R. & Swazey, J. (2017) *Spare Parts: Organ Replacement in American Society.* Abingdon: Routledge.

Franco, M., et al. (2013) Population-wide weight loss and regain in relation to diabetes burden and cardiovascular mortality in Cuba 1980–2010: Repeated cross sectional surveys and ecological comparison of secular trends. *British Medical Journal.* 346, 1515.

Freedman, J. & Fraser, S. (1966) Compliance without pressure: The foot-in-the-door technique. *Journal of Personality and Social Psychology.* 4(2), 195.

French, J. & Raven, R. (1968) The bases of social power. In Cartwright, D. & Zander, A. (eds.) *Group Dynamics.* New York, NY: Harper & Row, pp. 259–269.

Garwood-Gowers, A. (1999) *Living Donor Organ Transplantation: Key Legal and Ethical Issues.* Aldershot: Ashgate.

Garwood-Gowers, A. (2013) Time to address the problem of post-mortem procurement of organs for transplantation occurring without proper pre-mortem consent. *European Journal of Health Law.* 20, 383–408.

Garwood-Gowers, A. & Pereira, A. (2017) Human to human organ and tissue transplantation. In den Exter, A. (ed.) *European Health Law*. Leiden: Brill, pp. 391–418.

Goldstein, K. (1939) *The Organism*. New York, NY: American Books.

Gordon, E., et al. (2011) Informed consent and decision-making about adult-to-adult living donor liver transplantation (LDLT): A systematic review of empirical research. *Transplantation*. 92, 1285–1296.

Halevy, A. & Brody, B. (1993) Brain death: Reconciling definitions, criteria, and tests. *Annals of Internal Medicine*. 119(6), 519–525.

Harris, J. (2003) Organ procurement: Dead interests, living needs. *Journal of Medical Ethics*. 29, 130–134.

Hatzinger, M., et al. (2016) The history of transplantation. *Urologe A*. 55(10), 1353–1359.

Henderson, M. & Gross, J. (2017) Living organ donation and informed consent in the United States: Strategies to improve the process. *The Journal of Law, Medicine and Ethics*. 45(1), 66–76. Available at https://doi.org/10.1177/1073110517703101.

Hoffenburg, R. (2001) Christiaan Barnard: His first transplants and their impact on concepts of death. *British Medical Journal*. 323(7327), 1478–1480.

Janis, I. & Mann, L. (1977) *Decision-Making: A Psychological Analysis of Conflict, Choice and Commitment*. New York, NY: The Free Press.

Jarl, J. & Gerdtham, U-G. (2011) Economic evaluations of organ transplantations – A systematic literature review. *Nordic Journal of Health Economics*. 1, 64–67.

Jonas, H. (1974) *Philosophical Essays: From Ancient Creed to Technological Man*. Englewood Cliffs, NJ: Prentice-Hall.

Knoll, G. & Nichol, G. (2003) Dialysis, kidney transplantation, or pancreas transplantation for patients with diabetes mellitus and renal failure: A decision analysis of treatment options. *Journal of American Society of Nephrology*. 14(2), 500–515.

Knowler, W. C., et al. (2002) Reduction in the incidence of type 2 diabetes with lifestyle intervention or metformin. *New England Journal of Medicine*. 346, 393–403.

Kortram, K., et al. (2016) Towards a standardized informed consent procedure for live donor nephrectomy: What do surgeons tell their donors? *International Journal of Surgery*. 32, 83–88.

Kousoulas, L., et al. (2011) Living donor liver transplantation: Effect of the type of liver graft donation on donor mortality and morbidity. *Transplant International*. 24, 251–258.

Lennerling, A., et al. (2003) Becoming a living kidney donor. *Transplantation*. 76, 1243–1247.

Lentine, K., et al. (2012) Risks and outcomes of living donation. *Advance Chronic Kidney Disease*. 19, 220–228.

Lentine, K., et al. (2016) Perioperative complications after living kidney donation: A national study. *American Journal of Transplantation*. 16(6), 1848–1857.

Levitt, M. (2011) Educating the public to encourage organ donation. In Farrell, A-M., et al. (eds.) *Organ Shortage: Ethics, Law and Pragmatism*. Cambridge: Cambridge University Press, pp. 52–66.

López-Fraga, M., et al. (2014) A needed convention against trafficking in human organs. *The Lancet*. 383(9936), 2187–2189.

Lopp, L. (2013a) Best practice proposal: Legal safeguards for living organ donation in Europe in consideration of the current national regulations. In Ambagtsheer, F. & Weimar, W. (eds.) *The EULOD Project Living Organ Donation in Europe: Results and Recommendations*. Lengerich: Pabst, pp. 107–145.

Lopp, L. (2013b) *Regulations Regarding Living Organ Donation in Europe*. Berlin; Heidelberg: Springer.

Lopp, L. (2014) Legal issues in living donation in the member states of the EU. In Weimar, W., et al. (eds.) *Organ Transplantation: Ethical, Legal and Psychological Aspects*. Lengerich: Pabst, pp. 160–166.

Machado, C. (2005) The first organ transplant from a brain-dead donor. *Neurology*. 64(11), 1938–1942.

Maleck, W., et al. (1998) Unexpected return of spontaneous circulation after cessation of resuscitation (Lazarus phenomenon). *Resuscitation*. 39, 125–128.

Medawar, P. (1944) The behavior and fate of skin autografts and skin homografts in rabbits. *Journal of Anatomy*. 78, 176.

Midgley, M. (2000) Biotechnology and monstrosity. Why we should pay attention to the 'yuk factor'. *Hastings Centre Report*. 30(5), 7–15.

Miles, S. (1999) Death in a technological and pluralistic culture. In Youngner, S., et al. (eds.) *The Definition of Death: Contemporary Controversies*. Baltimore: John Hopkins University Press, p. 310.

Miller, F. & Truog, R. (2012) *Death, Dying and Organ Transplantation: Reconstructing Medical Ethics at the End of Life*. Oxford: Oxford University Press.

Mjøen, G., et al. (2009) Morbidity and mortality in 1022 consecutive living donor nephrectomies: Benefits of a living donor registry. *Transplantation*. 88, 1273–1279.

Mollaret, P. & Goulon, M. (1959) The depassed coma. *Revue Neurologique* (Paris). 101, 3–15.

Muntean, A. & Lucan, M. (2013) Immunosuppression in kidney transplantation. *Clujul Medical*. 86(2), 177–180.

Murray, J., et al. (1955) Renal homotransplantation in identical twins. *Surgical Forum*. 6, 432–436.

NHS England. (2018) Very low calorie diets part of NHS action to tackle growing obesity and Type 2 diabetes epidemic. Available at www.england.nhs.uk/2018/11/very-low-calorie-diets-part-of-nhs-action-to-tackle-growing-obesity-and-type-2-diabetes-epidemic. Accessed 8.3.2019.

Niederkorn, J. (2014) Corneal transplantation and immune privilege. *International Reviews of Immunology*. 32(1) Available at 10.3109/08830185.2012.737877.

Norton, L., et al. (2017) Electroencephalographic recordings during withdrawal of life-sustaining therapy until 30 minutes after declaration of death. *Canadian Journal of Neurological Sciences*. 44, 139–145.

OUP. (2019) Oxford English Dictionary definition of death. *Oxford Dictionaries Online*. Available at www.oxforddictionaries.com/definition/english/death. Accessed 8.3.2019.

Pallis, C. & Harley, D. (2009) *ABC of Brain Stem Death*. London: BMJ Books.

Pan, X. R., et al. (1997) Effects of diet and exercise in preventing NIDDM in people with impaired glucose tolerance. The Da Qing IGT and Diabetes Study. *Diabetes Care*. 20, 537–544.

Pascalev, A., et al. (2013) Expanding living donation in Europe: Attitudes, barriers and opportunities, results from a multi-country focus group study. In Ambagtsheer, F. & Weimar, W. (eds.) *The EULOD Project Living Organ Donation in Europe: Results and Recommendations*. Lengerich: Pabst, pp. 27–52.

Pearn, J. & Franklin, R. (2012) 'The impulse to rescue': Rescue altruism and the challenge of saving the rescuer. *International Journal of Aquatic Research and Education*. 6(4), 325–335.

Petrie, D., et al. (2001) Factors affecting survival after prehospital asystolic cardiac arrest in a basic life support-defibrillation system. *Canadian Journal of Emergency Medicine*. 3, 186–192.

President's Council on Bioethics. (2008) *Controversies in the Determination of Death*. Washington, DC: Government Printing Office. Available at https://repository.library.georgetown.edu/handle/10822/559343. Accessed 19.2.2019.

Price, D. (1997) Organ transplant initiatives: The twilight zone. *Journal of Medical Ethics*. 23, 170–175.

Price, D. (2000) *Legal and Ethical Aspects of Organ Transplantation*. Cambridge: Cambridge University Press.

Ramachandran, A., et al. (2006) The Indian Diabetes Prevention Programme shows that lifestyle modification and metformin prevent type 2 diabetes in Asian Indian subjects with impaired glucose tolerance (IDPP-1). *Diabetologia*. 49, 289–297.

Reemtsma, K. (1995) Xenotransplantation: A historical perspective. *ILAR Journal*. 37(1), 9–12.

Sami, W. (2017) Effect of diet on Type 2 diabetes mellitus: A review. *International Journal of Health Sciences (Qassim)*. 11(2), 65–71.

Shah, S. K., et al. (2011) Death and legal fictions. *Journal of Medical Ethics*. 37(12), 719–722.

Shanteau, J. (1992) Competence in experts: The role of task characteristics. *Organizational Behavior and Human Decision Processes*. 53, 252–266.

Shewmon, A. (1998) 'Brainstem death', 'brain death' and death: A critical re-evaluation of the purported equivalence. *Issues in Law and Medicine*. 14(2), 125–145.

Shimazono, Y. (2007) The state of the international organ trade: A provisional picture based on integration of available information. *Bulletin of the World Health organization*. Available at www.who.int/bulletin/volumes/85/12/06-039370/en/. Accessed 8.3.2019.

Simmons, R., et al. (1977) *The Gift of Life*. Chichester: John Wiley & Sons.

Somjen, G. (2004) Irreversible hypoxic (ischemic) neuron injury. In Somjen, G. (ed.) *Ions in the Brain*. New York, NY: Oxford University Press, pp. 338–372.

Spital, A. (2004) Donor benefit is the key to justified living organ donation. *Cambridge Quarterly of Healthcare Ethics*. 13(1), 105–109.

Spital, A. (2006) Living organ donation: Striking a proper balance and obtaining valid consent. *Transplantation*. 82(9), 1244–1245.

Spital, A. & Jacobs, C. (2007) The beauty of the gift: The wonder of living organ donation. *Clinical Transplantation*. 21(4), 435–440.

Spital, A. & Taylor, J. (2008) Primum non nocere and living organ donation: A response to Dr. Klintmalm. *American Journal of Transplantation*. 8(6), 1352–1353.

Stothers, L., et al. (2005) Attitudes and predictive factors for live kidney donation: A comparison of live kidney donors versus nondonors. *Kidney International*. 67(3), 1105–1111.

Tedesco, D. & Haragsim, L. (2012) Cyclosporine: A review. *Journal of Transplantation*. Available at www.hindawi.com/journals/jtrans/2012/230386/. Accessed 8.3.2019.

Thukral, S., et al. (2018) Long term consequences of complex living renal donation: Is it Safe? *Transplantation Proceedings*. 50(10), 3185–3191.

Tiessen, J., et al. (2008) *Improving Organ Donation and Transplantation in the European Union, Rand Europe*. Available at www.rand.org/content/dam/rand/pubs/technical_reports/2008/RAND_TR602.pdf. Accessed 8.3.2019.

Townsend, C., et al. (2016) *Sabiston Textbook of Surgery*. Amsterdam: Elsevier.
Tuomilehto, J., et al. (2001) Prevention of type 2 diabetes mellitus by changes in lifestyle among subjects with impaired glucose tolerance. *New England Journal of Medicine*. 344, 1343–50.
Unger, E. (1909) Nierentransplantation. *Berl Klin Wochenschr*. 1, 1057.
Veatch, R. (1999) The conscience clause. In Youngner, S., et al. (eds.) *The Definition of Death: Contemporary Controversies*. Baltimore: John Hopkins University Press, p. 310.
Veatch, R. (2005) The death of whole-brain death: The plague of the disaggregators, somaticists, and mentalists. *Journal of Medicine and Philosophy*. 30(4), 353–378.
Veatch, R. (2008) Donating hearts after cardiac death – Reversing the irreversible. *New England Journal of Medicine*. 359, 672–673.
Weisfeldt, M. & Becker, L. (2002) 'Resuscitation after cardiac arrest' A 3 – Phase time-Sensitive model. *Journal of American Medical Association*. 288(23), 3035–3038.
Wertheimer, P., Jouvet, M. & Descotes, J. (1959) Diagnosis of death of the nervous system in comas with respiratory arrest treated by artificial respiration. *Presse Medicale*. 67, 87–88.
Westlie, L., et al. (1993) Quality of life in Norwegian kidney donors. *Nephrology Dialysis Transplantation*. 8, 1146–1150.
Wetmore, J. & Collins, A. (2016) Global challenges posed by the growth of end stage renal failure. *Renal Replacement Therapy*. 2, 15.
WHO/ONT. (2019) *Global Observatory on Donation and Transplantation*. Available at ww.transplant-obervatory.org. Accessed 15.1.2019.
Williams, N. (2018) On harm thresholds and living organ donation: Must the living donor benefit, on balance, from his donation? *Medical Health Care and Philosophy*. 21(1), 11–22.
Zirm, E. (1906) Eine erfolgreiche totale Keratoplastik. *Graefe's Archive for Ophthalmology*. 64, 580–593.

6 Human tissue research

6.1. Introduction

Conducting research on body parts and tissues is nothing new – having always been a part of anatomical examination which is known to have been practiced since at least ancient Egyptian times (Lassek, 1958). However, over the course of the modern era in general and the twentieth century in particular, the process of extracting, storing and using materials from the body for research purposes has become ever more systematic and scaled (Hoeyer, 2005), reflecting and advancing the way the character of medicine has shifted and more recently the growth of scientific and economic interest in the biomedical technosciences (Bister, 2011, 169). Nowadays, a vast and ever increasing amount of financial resources are now devoted to it by governments and industry across the globe (Vaught et al., 2011). Small and large research biobanks have proliferated (Henderson et al., 2013); population biobanks and cohort studies have been developed; and many smaller collections have been consolidated under broader research infrastructure, coinciding with considerable investment in sequencing technology and the funding of research consortia and disease networks with specific research objectives (Chalmers et al., 2016).

As a result, human tissue research has become highly embedded in many societies. It has also come to be regarded by many commentators as integral to the advance of medicine. Weir and Olick (2004, 3), for example, suggest that it is one of the three legs on which modern medical progress stands – the other two in their view being research with non-human animals and research in chemistry and pharmaceuticals. However, I argue in this chapter that human tissue research is often more problematic in the ends it pursues than it first appears. The way in which it is being relied on is reflecting and reinforcing certain problematic trends, including late stage capitalism and the move toward mechanistic and industrialized systems of medicine. Much of it can be seen to fall short of being productive as a result. Furthermore, even leaving questions of its productivity to one side, tissue research is being regulated and practiced in ways that are incompatible with the worth of those used.

6.2. Nomenclature issues

Most instances of tissue use both for research and other purposes are of necessity preceded by a period of storage. Much of the time the storage is together with other material. What constitutes the appropriate nomenclature for such collections is contested, with a range of overlapping and contrasting terms being used in practice (see Hoeyer, 2008, 429). Terms like 'biorepository' and 'tissue repository' would be appropriate if one were only referring to storage facilities, but most of the time, there a range of broader activities that are occurring. These include things like organization, processing, distribution and practical use of samples and data relating to them.

The term 'biobank' is more apt in this regard since a 'bank' suggests an entity organized toward certain ends and seeking to fulfil these on a long term or permanent basis via the range of things it does with the materials and related data that it takes in. De Souza and Greenspan's (2013) definition of a biobank as 'a biorepository that accepts, processes, stores and distributes biospecimens and associated data for use in research and clinical care' captures this as does Meir et al.'s (2014) definition of it as entailing collections of biological samples and associated data that are systematically organized for use by stakeholders, such as researchers and healthcare providers. Both definitions are technically correct to assert that biobanks have enabled the activities and uses that they refer to, but it is important to also stress that they have typically not done so in an unconstrained way.

However, it is also true that the practice of biobanking has in the past been allowed to run ahead of academic discourse – let alone public debate – about how it should be constrained. This is reflected in empirical analysis conducted on references to biobanking ethics within publications locatable on the Web of Science and PubMed databases. Of those published prior to November 2010, over half were specifically published in 2009 or 2010 (Budmir et al., 2011). This is not because the issues being raised are new either – the specific issues most commonly raised were informed consent, privacy and identifiability, return of results to participants, importance of public trust, involvement of children, commercialization, the role of ethics boards, international data exchange, ownership of samples and benefit sharing (Budmir et al., 2011), all of which have been around since biobanking began.

One can argue that at least most of these issues are also relevant to human tissue research more generally. Nonetheless, there are at least two reasons why biobanking should be regarded as more than simply a 'sexed up' descriptor for human tissue research. The first is that, whilst some definitions imply that it is confined to the research context (e.g. Oxford University Dictionary). it is clearly also being employed in other contexts as well, such as that of the mass storage of materials like bone marrow and blood pursuant to their prospective use in transplant therapy. The second is that even biobanking in the research context – or research biobanking as Hoeyer calls it (2008, 429) – is difficult to treat as a synonym for human tissue research. Partly this is because it is often used more selectively in practice. Indeed, it is often 'hierarchically reserved' for those research focused 'collections' of tissue that satisfy certain preconditions and requirements as to their organization and governance (Hoeyer, 2008, 429). Some of these preconditions and requirements are normative as between disciplines and jurisdictions, but others vary (Lenk et al., 2011, X) rendering it impossible to define biobanking in a universally acceptable way.

6.3. The relevance of key overarching approaches to ethics

Discourse concerned with the ethics of human tissue research has been characterized as centred on the need to safeguard donor autonomy (Bister, 2011, 170) and also as reflecting a tension between such autonomy and the common good which is 'seen to have particular salience' in this context because of the dependence of researchers upon entire communities (including healthy populations), many of whom are unlikely to benefit from the results of the research' (Lipworth et al., 2011). However, it is actually more eclectic than this.

Principles not easily reducible to either autonomy or the common good that are frequently given attention in it include solidarity, reciprocity, trust and protection of vulnerability. And some thinkers have supported an overarching approach to use in this field that is more protective than an emphasis on either autonomy or the common good would imply. Barilan, for example, has emphasized dignity and rights (2011, 3–14) and in the course of doing so stressed, amongst other things, the problem of excessive alienation of interests that can result from an emphasis on autonomy and how it can manifest as exploitation in the context of biomedical use of the body generally and human tissue research specifically. Meanwhile, Van Assche and Sterckx have emphasized dignity (2014, 265–287) with a strong slant on the beyond the individual dimensions of its protection that are not easily reached via autonomy. I have not dissimilarly emphasized respect in this context in a way that associates it with protecting all members of the human species equally and avoids instrumental use of any (Garwood-Gowers, 2011, 25–33). In Chapter 1 of this book, I built on that further by arguing that respect for human worth in the context of medical use requires that use to be productive in the end it pursues and consistent with four norms in the way that it pursues them, namely:

(1) protecting all members of the species as of equal inherent value;
(2) protecting human nature and form;
(3) protecting the species and its members from instrumentalization; *and*
(4) appropriate account being given to any relevant non-absolute principles, such as autonomy and beneficence.

It is clear that this emphasis is incompatible with an absolutist approach to autonomy, the common good or any hybrid mix of the two that might be dreamed up. Absolute protection of autonomy opens the door to human tissue research that unacceptably damages the non-autonomy related interests of humans beings by being unproductive, counterproductive or excessively detrimental and human tissue research that exploits the particularly vulnerable and/or communities and/or denigrates human beings as a species. Meanwhile appeals to the 'common good' and similar in human tissue research are little more than blunt tools to propagate its expansion at the cost of respect for worth. In a milieu that is largely uncritical of the medical orthodoxy and vested commercial and career interests that underpin the hue of such research, demand for it is too easily assumed to represent a need for it. Acceptance of a (false) dichotomy between the interests of the individual and the whole then sets the scene for such demand to be used as a lever for carrying out research even when it falls short of one or more of the aforementioned norms. Indeed, it seems that practically whenever there is a public reflection on what constitutes appropriate constraint in this area numerous individuals and bodies come 'out of the woodwork' to make the case for the protection of the individual to be 'weighed against' the common good.

6.4. Human tissue research and productivity

The extensively propagated notion that human tissue research is a highly productive activity has set the scene for the easier justification of both the high levels of use of financial resources and human beings that have been required to bring it about on a vast scale. However, closer analysis reveals that there are good reasons to question generalized assertions of the value of such research. One of these reasons is that there is a suspicion, supported by some empirical evidence, that such research generally and biobanking, in particular, tends to be represented in a 'hyped' way with its near future benefits exaggerated and risks and limitations minimized, not least within the popular press (Ogbogu et al., 2014). Another is that its value is open to question even when we simply consider the ends that it is aimed at. Some of these ends are inherently questionable – such as enabling the creation of biotechnological products. Meanwhile, many more are capable of being constructive or destructive – such as attempting to improve understanding of the nature of specific tissues and the health effects that specific actions or agents can have on them and the genetic make-up, background and health of those they are removed from and potentially their kin, the wider community they hail from and even humanity as a whole. Furthermore, whilst some are unequivocally constructive in principle – such as attempting to advance the potential of personalized care and also to bring improvements to how we diagnose/treat/cure both common and rare diseases – even their pursuit does not automatically produce value.

Furthermore, there are fundamental issues with the way that human tissue research is generally oriented and structured which bear upon its productivity. One of these is that most of it reflects and reinforces the counterproductive way in which human health is handled in many modern societies. As I concluded toward the end of Chapter 1, modern industrialized societies have: failed to maintain the needed predominant focus on the environmental, social, economic and cultural conditions that were critical to the promotion of health; allowed medical systems to become a central feature in spite of their more marginal capacity to positively impact on life and tendency toward counterproductivity; *and* have specifically allowed them to take on and maintain a counterproductive mechanistic orientation.

A shift toward productivity would therefore centre on restoring the role of society in health, radically downsizing and limiting the role of the medical system and reorienting it toward being holistic. Human tissue research specifically would become much more marginal in scope and scale than it currently is and much more constrained in the ends which it pursues. Whilst there are no significant signs that this kind of level of constraint is about to manifest, there are at least some signs of a greater willingness to scrutinize its value more closely. In the context of population biobanks, for example, funders have begun to question 'their commitment to providing support in perpetuity,' and some countries, including Australia and Singapore, have already withdrawn their funding (Chalmers et al., 2016).

Another issue with such research is that its social organization reflects and reinforces problems within society more generally. There has been considerable effort put into preventing profit being derived, as such, from the body as a whole and materials taken from it but no real effort put into preventing profit being made from research on either. That lack of effort is even evident when the profit derives from the research-based development of biotechnological products which themselves ultimately literally derive from materials taken from the body. This has dramatically added to both the focus on the use value of the human body and the more insidious use of it as a tool to be commercially exploited. Tissue-based economies have emerged, with a large number of commercially driven biomedical companies lining up a vast number of biomedical researchers to prospect the body for an array of materials that they can analyze and transform with a view to developing money spinning products. This is an industry which even just by 2001 in the US alone comprised of 1,300 companies and was worth $17 billion USD (Andrews & Nelkin, 2001, 20). By 2010, the global market had reached $141 billion USD and was projected to increase to an estimated $183 billion USD by 2015 (BCC, 2011).

This late capitalist commodification or 'marketization' of the human body can be associated with a range of problems. Lipworth et al. (2011) have observed how critiques have focused on its ability to adversely impact the sanctity of the body, human psychological development, dignity, well-being, agency and kinship. The way that tissue research tends to be organized is not only enabling these adverse impacts to be felt on it but also having such ill effects as crimping the potential for co-operation and co-ordination between different sets of researchers and enhancing levels of proliferation and duplication of research. At the same time, these impacts could be moderated if states regulated science in such a way as to reduce commercially driven opportunism and exploitation (Hoeyer, 2005; Waldby & Mitchell, 2006). This goal should be seen as extending to preventing tissue and related data collected by not-for-profit organizations from being exploited for commercial purposes by other organizations. Ultimately, ways must be found to make sure the use of tissue and related data does not violate social interests.

6.5. Shortfalls in respect for the individual with regard to the justification for drawing on their material and related data for research

According to an ethic of protecting human worth and to related general and human rights based legal norms in the health field, individuals should only have their body material and related data drawn on in the tissue research context where that is justifiable in terms of their interests. However, this imperative has been significantly deviated from in a number of important respects in tissue-specific practice, governance and even law. Perhaps the most important of these is that the protective rules that should be – and generally are – associated with the principle of consent have to some extent been pushed to one side by distortions in the framing and use of it in the tissue research context. The most central of these rules is that consent must meet four requirements to be valid. Specifically, it must be: (1) given to an intervention that is of a type that is capable of being validly consented to, (2) by someone who has the legal power to give it and in a manner that is sufficiently (3) free *and* (4) informed. It is the third and fourth requirements, in particular, that have been handled problematically in the human tissue research context. Before examining how it is important to reflect on what these requirements are trying to achieve.

In human rights terms, they are designed to ensure that the decision-making process sufficiently protects the interests of prospective subjects of medical intervention having regard to the particular circumstances of the case and the nature of the decisions to be taken. This is a matter of autonomy and hence in ECHR terms naturally falls within the scope of the Art 8(1) right to private life as emphasized in the case of *V.C. v Slovakia* App no 18968/07 (ECtHR, 8 November 2011, para. 141). The Applicant in this case, had been subject to sterilization after she had signed a consent form for it to be performed but had not been fully informed about her health status, the proposed procedure or alternatives to it and had been asked to consent while she was in labour, shortly before the performance of a Caesarean section and without having had an opportunity to consider and reflect on the issues and discuss them with her partner. The Court took the view that the pressure her ability to exercise free will had been undermined not just to the effect of violating her right to private life but also to the effect of being inhuman and degrading treatment and thus in violation of Art. 3. From this, one can abstract that at least when a decision on a serious matter is being made, the decision-maker must be both equipped by disclosure of the key information pertaining to it and left with space and time to consider that information, consult with relevant others and ultimately make up their own mind what to do.

The information aspect was given further scrutiny in *Csoma v Romania* App no 8759/05 (ECtHR, 15 January 2013) where the Court held that, pursuant to the Art. 8(1) right to private life, contracting states are bound

> to adopt necessary regulatory measures to ensure that doctors consider the foreseeable consequences of a planned medical procedure on their patients'

> physical integrity and to inform patients of these consequences beforehand, in such a way as that the latter are able to give informed consent.
>
> (para. 42)

Similarly, CHRB Art 5. requires 'appropriate information' on not just the nature and purpose of proposed interventions but also their consequences and risks to be given before they are carried out, and its Explanatory Report requires the latter to be reasonably well tailored to the particularities of the prospective subject's position (for example, accounting for special risks posed due to pathologies) and adequate answers to be given to requests for additional information (para. 35). It has been suggested that it is a standard that requires not just information but also explanation to be given on such matters as:

> How the intervention is being performed; where it consists of a process of interventions over time, the nature of what is entailed in that process, its potential length etc.; what, if anything, is expected of the prospective subject of the intervention or process of interventions; progress made; and its prospective short and long term benefits and harms and risk along with those of alternative courses of action, which should also be compared.
>
> (Garwood-Gowers & Olsena, 2017, 263)

It is notable that some European, as well as some non-European jurisdictions, do not generally go down such a track of requiring consent to be properly informed. Under the common law, for example, consent to medical intervention is generally treated as real in informational terms where the person empowered to give it has been informed of its broad nature (per Bristow J in *Chatterton v Gerson* [1981] QB 432 at 442). This rule reflected a deliberate effort to reserve civil and criminal trespass to situations in which medical professionals had engaged in egregious wrongdoing, providing recourse for more ordinary failures via less stigmatizing forms of legal action. The most commonly relevant of these has been the tort of negligence which started out somewhat unfit to play this role but has ended up being relatively effective at doing so due to technically sound developments in its approach to breach of duty and rather more questionable developments in its approach to factual causation (see further Garwood-Gowers & Olsena, 2017, 264–265). Also of note is that the rule laid out in *Chatterton v Gerson* has exceptions relevant to the immediate context. In particular, there is authority for the proposition that informed consent is required in the non-therapeutic research context (*Halushka v University of Saskatchewan* (1965) 53 DLR (2d) 436); by logical extension informed consent should be required for any intervention (i.e. research or otherwise) that deviates from optimal protection of the clinical interests of the prospective subject of it (Garwood-Gowers, 2005); *and* many jurisdictions require informed consent with respect to participation in clinical trials (EU member states, for example, are required to do so under EU law – first the Clinical Trials Directive 2001/20/EC and now the instrument that repealed it – namely, Regulation (EU) No 536/2014 on clinical trials on medicinal products for human use).

The obligation to inform people of the nature, purpose, implications and alternatives to interventions that they are being asked to agree to may be normative, but it is also heavily resisted by most stakeholders in the human tissue research field specifically. Empirical research confirms the commonly received wisdom that this is particularly so in the research biobanking context (Garrison et al., 2015). Indeed, in this context, prospective participants have even often been asked to sign up to a 'one time general consent' to future use of their removed material for a broad research purpose or set of purposes (see Figure 6.1 for the UK Biobank as an exemplar).

Whilst some thinkers have endorsed broad consent (van der Graaf & van Delden, 2011; Wendler, 2006), it attracts a great deal of public opposition (De Vries et al., 2016; Murphy et al., 2009; Simon et al., 2011; Garrison et al., 2015). It has been rightly commented that it does not seem to fulfil legal and ethical informational requirements (Allen et al., 2013; see also Caulfield & Murdoch, 2017). It runs counter to personal sovereignty in a very fundamental way since it entails asking individuals to give up control of some of their tissue in perpetuity, essentially on the basis of (naïvely) trusting that it will be used productively (see concerns in Section 6.4 earlier) and compatibly with their current and future objectives and values.

The UK Bio-bank was set up to recruit half a million 40–69 year old people from across the UK with the very wide-ranging aim of 'improving the prevention, diagnosis and treatment of a wide range of serious and life-threatening illnesses – including cancer, heart diseases, diabetes, arthritis and forms of dementia' (www.ukbiobank.ac.uk/about/what.php). Those being recruited are asked, amongst other things, to agree to permit: Access to their medical and other health-related records and for long-term storage and use of this and other information about them, for health-related research purposes (even after my incapacity or death); long-term storage and use of their blood and urine samples for health-related research purposes (even after their incapacity or death); and to relinquish all rights to the samples which they donate (www.ukbiobank.ac.uk/docs/Consent.pdf).

Since none of the results are given to participants (except for some measurements during this visit) they are effectively being asked to hand over genetic material and relinquish the privacy of data to an organisation without getting any direct health benefit in return.

Figure 6.1 UK Biobank's model of consent

Despite this, it has been largely embraced in the US. The latest version of the Federal Policy for the Protection of Human Subjects, created in 2017, permits it for storage, maintenance and secondary research use of even identifiable private information and identifiable bio-specimens. Furthermore, since 2014, the National Institutes of Health have made funding of most forms of genomic research conditional on resort to it both for research and data sharing (NIH, 2014). By contrast, within member states of the EU and the wider European Economic Area, the scope for it has been slightly contained via the new General Data Protection Regulation (EU) 2016/679. This builds on the right of protection of personal data afforded to natural persons under Art. 8(1) of the Charter of Fundamental Rights of the European Union (the 'Charter') and Art. 16(1) of the Treaty on the Functioning of the European Union (TFEU). It stipulates that consent to data processing should be given by a clear affirmative act establishing a freely given, specific, informed and unambiguous indication of the data subject's agreement to the processing of personal data relating to him or her (reg 32). Silence, inactivity and not unclicking pre-ticked boxes are specifically referred to as not constituting consent (reg 32). In relation to processing data in scientific research contexts where the purpose of doing so is not fully identified, it stipulates that 'data subjects should be allowed to give their consent to certain areas of scientific research when in keeping with recognised ethical standards for scientific research' (reg 33) and goes on to stress that they should have the opportunity to tailor it to only certain areas of research or parts of research projects to the extent allowed by the intended purpose (reg 33).

The ethical damage that has been caused by tolerating broad approaches to consent in the past could be mitigated to some degree if a policy of requiring consent to new uses of stored samples was adopted. However, whenever such a policy has been considered, it has often been rejected for reasons of expedience. The NHANES III Project is a good example. This was the last of a three-part series of medical and epidemiological studies carried out in the US by the Centers for Disease Control and Prevention (CDC). Ran from 1988 to 1994, it involved 40,000 participants of whom 29,314 had samples taken from them under a simple general consent for storage and testing but without any more specific information or consent. By the mid-1990s, these samples were being used for molecular research with DNA without any attempt having being made to recontact participants and gain their specific consent to this – that solution having been explicitly rejected on the grounds that doing so would cost $2 million USD. Whilst empirical data suggests that historically most people have not been interested in being recontacted (Hansson, 2005; Wendler, 2006), this only goes to support the hypothesis that there has been too much complacency around the purposes for which human tissue research may be used and, relatedly, the values it may express and encourage. There is no guarantee that stored tissue will not be used to conduct clearly unethical research or that data from its use will not be used for unethical purposes or simply that future uses will not clash with the existing or future interests of those who are used, such as by clashing with their existing or future values. This is undoubtedly especially so in our era of genetic research which, as a matter of course, will uncover a great deal of especially sensitive information about those whose tissue is used and those to whom they are genetically closely or even distantly related to and thus present potential problems in terms of their dignity (see further Kirchhoffer & Dierickz, 2012).

Broad consent is also rendered more problematic when one places in context the fact that those who have given their material are unlikely to enquire as to the subsequent uses of it and thus unlikely to consider withdrawing their consent. In spite of this, some discourse has gone as far as arguing that the right to withdraw – which is a key element of consent law generally and the governance of medical research specifically – should be curtailed in this context. For example, Eriksson and Helgesson (2005, 1076) argue that withdrawal should only be accepted where the reasons for it are both significant and cogent. This is a quite astonishing view, treating those who have chosen to contribute to human tissue research as like flies that once trapped on sticky paper are only to be released where others deem their reasons for release to be good enough. It is an affront to their freedom that is quite blatantly motivated by the desire to maximize their use. Whilst it is true that withdrawal of consent can impede research quality, the right approach in terms of human worth or dignity is quite clearly for researchers to build the possibility of it into their calculations of whether the research is worthwhile trying to carry out in the first place.

It is also notable that standards of protection with respect to tissue removal are often diluted even further with respect to certain more vulnerable classes of individual. The Human Tissue Act 2004 scheme for England, Wales and Northern Ireland provides a good case study of how this dilution can operate. The default position under this scheme is that removal is subject to a requirement of 'appropriate consent' (section 1). Whilst this default generally applies to the dead (section 1(4)), it is abandoned when the removal is from an imported body and also abandoned when it is from a person who died at least a hundred years before the law came into place (Human Tissue Act section 1(5) and 1(6) read in conjunction with section 1(4), section 1(1)c and schedule 1). Furthermore, when it does apply, it is treated as satisfied not only in cases where the deceased gave consent pre-mortem but also where a relevant member of kin has given proxy consent post-mortem with the proviso that the deceased did not veto removal during their lifetime (section 3(6)).

This is an ethically dubious reliance on proxy consent since the only interests a person retains after they die lie in the protection of their bodily integrity, wishes, values and beliefs – all of which point to leaving them free of medical use that they have themselves not consented to prior to death in all situations except where they themselves appointed someone to decide the matter for them. Similarly, dubious is the fact that proxy consent can be used to justify removal for research purposes from competent children in situations where they have not voiced objection to this (Human Tissue Act section 2(3)). The Act also allows for embryos outside the body to be subject to DNA analysis without qualifying consent (section 45(2)(c)), and it is also notable that under the separate legislative scheme of the Human Fertilisation and Embryology Act 1990, certain forms of research are allowed to be carried out on early stage embryos under licence (see particularly section 11(1)(c) in conjunction with schedule 2, para. 3). Many other jurisdictions also allow some research on early stage embryo. The stipulation within Art. 18(1) of the CHRB that in-vitro embryos should be adequately protected in the research context has proved to be an ineffective barrier to this, though at least most jurisdictions have diverged from England and Wales in following its Art. 18(2) stipulation that they should not be created for research purposes.

It is not just where it is being justified by reference to consent, but also otherwise, that the purpose(s), scope and implications of a removal of tissue ought to be transparent and specific. An attempt to bring about use beyond these criteria should be classified as secondary use and as requiring its own justification by reference to consent or otherwise, with its purpose(s), scope and implications also needing to be transparent and specific. In the research context, secondary use often entails researchers expanding on a non-therapeutic research purpose or set of purposes that predicated tissue being removed from the individual(s) concerned in the first place. However, equally, it often entails non-therapeutic research being carried out on material that was removed from them for an entirely different reason – such as to satisfy an obligation of law (e.g. pursuant to the detection of the cause of a death) or to provide for their health via diagnosis and/or treatment (e.g. where cancer is suspected).

There are no global figures for how frequently secondary use as a whole takes place, but one may speculate that there are perhaps hundreds of millions of incidences of it every year. In this light, ensuring that it is carefully constrained by reference to respect is vital. My position in this regard starts with the notion that one should not trivialize the nexus that materials has with us merely because it has already been removed. I would not describe it as belonging *to* us – not least because of the possessive and even property based implications of such a description. However, spiritually and/or intuitively it is easy to reach the conclusion that it belongs *with* us. After all if, for example, one had part of a finger go missing in an ice skating incident, one would still see it as one's finger and be aggrieved at the perceived sleight to one's rights were someone else to make off with it. The same would also be true if a nurse syphoned off one's blood during its passage through a dialysis machine. Even when removed, the material has a symbolic and intensely personal significance. Furthermore, in the modern era, it has begun to carry additional significances based on the added possibilities for its use. There are, for example, added possibilities of it being used in ways that might conflict with productive value (see earlier section) and/or with what those individual(s) subject to it holds to be of value. There are also added possibilities for it to be used in ways that can uncover highly personal or otherwise significant data about those being used, the genetic community they hail from and humanity as a whole.

Discourse, practice and to some extent even governance and law specific to secondary use clashes with this view in several ways. Part of the problem is that use for research purposes following removal for other purposes is often treated as if it is merely a useful repurposing of what would otherwise be wasted. Indeed, material in this category is regularly called 'leftovers' – a description which casts researchers using it as kind of a scientific equivalent of *The Wombles* – environmentally friendly teddies lyrically described by their 'biographer' as spending part of their days 'making the use of the things that they find, things that the everyday folks leave behind' (Beresford, 1968). However, the idea that body materials can simply be treated as abandoned and freely repurposed (The Nuffield Council on Bioethics, 1995) is not only legally suspect (Matthews, 1995), but also morally dubious given their earlier outlined significance. Furthermore, whilst it has not become a template for secondary use, the cavalier attitude behind it and the related language of waste certainly to some degree reflect and are reflected in approaches to governance and law in this area.

Moore v Regents of University of California 793 Pd 479 (1990) – An illustration of the misuse of human beings that can occur when tissue removal and use is not properly policed and controlled.

The key facts of this case were that the plaintiff, John Moore, had been diagnosed with hairy cell leukaemia in 1976 and on the clinically appropriate advice of Dr Golde at the UCLA Medical Center of Los Angeles he had his spleen removed but was then also subsequently encouraged by Dr Golde to travel regularly to the Center over the next seven years from his home in Seattle. He was given by the impression by Dr Golde that these visits and the extractions of various samples that occurred during them were necessarily and solely designed to help ensure his continued health. In fact, Dr Golde and his associate Dr Quan were developing an immortal cell line from his material which by the time the case had ended up in the Supreme Court of California was estimated as having a potential commercial value of $3.01 billion. Ultimately, the Court accepted one of Moore's legal contentions - namely that the fiduciary duty owed to him gave rise to a duty to give information about planned use of his extracted material. The case illustrates at least three problems that relate to the scientific and commercial value placed on tissue:

Firstly, it can be an incentive for unscrupulous behaviour in which a removal of tissues is presented as being for treatment and/or diagnostic purposes when in fact those directing its removal know it is not. Such removals would fall foul of criminal, civil and human rights based laws protecting the person since the purported consent to them is vitiated by fraud. However, they are likely to occur from time to time in spite of this.

Secondly, it provides an incentive for lower level unethical behaviour in the form of removals that are warranted for treatment and/or diagnostic purposes being presented as if they are solely for such purposes when in fact there is an inkling or a plan that they may then be put to use to other ends.

Thirdly, it will inevitably lead to widespread commodification and the spectre of people profiting from their wrongdoing where there are no real constraints on people profiting from 'working' on material.

Figure 6.2 Moore misuse

Moore is in many respects an echo of the Henrietta Lacks case - involving a woman who had cancer whose removed material was used without her consent and became the basis of the HeLa cell line. Decades after her death some of the legal issues around use of her material are yet to be resolved. In both the Moore case and the Lacks case the cell lines were named after the individuals whose material was used for them, which can be read as an insulting and patronising attempt to placate them.

Figure 6.2 (Continued)

In terms of international health law instruments, the CHRB sets the tone in this regard. Whilst Art. 22 of it states on its face that appropriate information and consent procedures should be applied to storage and use of body material for a purpose other than that for which it was removed what these arrangements actually entail is fairly dilute. Specifically, para. 137 of the Explanatory Report stipulates that express consent should not be considered a systematic requirement but rather as necessary in certain circumstances. It suggests that the collection of sensitive information about identifiable individuals is one such circumstance, and it can be argued that this is a widely applicable circumstance since, as para. 135 of the Explanatory Report notes, 'much information on the individual may be derived from any part of the body, however small (for example blood, hair, bone, skin, organ)' and goes on to add that '(e)ven when the sample is anonymous the analysis may yield information about identity.' Nonetheless, the certain circumstances comment in para. 137 is anathema to the basic principle of applying informed consent that underpins Art. 22. This is made all the more clear by the fact that para. 137 specifically green lights the abandonment of consent in two circumstances.

The first of these is where the relevant decision-maker has been informed and does not object. This is sometimes described as 'presumed consent,' which is clearly an oxymoron since at a very basic level consent revolves around agreement and, consequently, is not something that can be presumed simply from lack of objection (Garwood-Gowers, 2013). It is a democratically deficient approach because rather than promoting and protecting everyone's enjoyment of their freedom from intrusion it circumscribes that freedom in order to try and meet demand for research. It also specifically has a more pronounced effect on those who do not resist its effects by objecting. These are often going to be those who are especially vulnerable 'such as many of the sick and the elderly, those with cognitive impairments, drug addicts and those who are unable, unwilling or simply unlikely to actively protect their interests' (Árnason, 2007). Such classes might reasonably expect that society would be particularly vigilant about protecting them because of their vulnerability rather than treating it as something to be exploited. Another problem with presumed consent that is specific to when it is used for data is that there is no way of guaranteeing that mined data will not be associated back to the individuals it 'belongs' to.

There are a number of reasons for this, all of which can be illustrated via the example of Iceland's Health Sector Database, which was set up under its Health Sector Database Act (No. 139/1998). This Act enabled medical records kept in hospitals and other healthcare institutions to be accessed by a licenced third party for the purpose of obtaining genetic data if the individuals to whom those records belonged did not object to that within a limited time frame. The Icelandic Government had already been negotiating with a third party in the run up to the legislation being passed. This third party was misleadingly presented to the population as being Icelandic whereas, in fact, it was actually the US company DeCode Genetics working through its Icelandic subsidiary (Árnason, 2007). DeCode Genetics used one-way encryption, but even this is not 100% secure, especially from the malfeasance (including potential mal-intent) of those working with the data.

Furthermore, the motivation to breach security – and thereby the potential for it to be breached – is higher where the data is potentially very valuable, such as where (as in this project) it is genetic data. The use of presumed consent in relation to data also leaves individuals whose data is used exposed to a risk of identification that they did not consent to running. Inevitably that risk is going to be a lot higher as well in settings where the data is known to be drawn from a relatively small, stable and very close-knit population such as the one in Iceland. It was rendered even higher still in this case by the decision to publish a rich range of data including 'age group (within a 5-year bracket), town or municipality of residence, education, marital status and, of course, all sorts of health information' (Árnason, 2007). The legislation ended up being challenged in the Courts by a woman whose deceased father's data was non-consensually included on the database. Ultimately, the Icelandic Supreme Court in *Gudmundsdóttir v the State of Iceland* (Judgment No. 151/2003) ruled it unconstitutional on the basis that it was insufficiently protective of the right to privacy guaranteed by Art. 73 of the Icelandic Constitution.

The second situation in which para. 137 greenlights the abandonment of consent is where it is not possible or very difficult to find the persons concerned again in order to ask for their consent. This is a theme that is also repeated in the CHRB's 2008 Additional Protocol concerning Genetic Testing for Health Purposes, Art. 14 of which leaves scope for signatory states to allow genetic tests to be performed on the previously removed biological materials of a person who cannot be contacted. The provisos to this are that the use must be the benefit of his or her family, proportionate, the expected benefit cannot be otherwise obtained and the test cannot be deferred. These may superficially seem to be rigorous constraints, but actually they do not alter the basic fact that the Article as a whole is structured in such a way as to leave signatories able to comply with it whilst at the same time perpetrating use that is at odds with respect.

This mentality of diluting protection in the context of secondary use of material and data derived from it has also been embraced at the domestic level. Within Europe, the position that domestic jurisdictions take as regards data is harmonized via the General Data Protection Regulation 2016/679 which builds on the right of protection of personal data afforded to natural persons under Art. 8(1) of the Charter of Fundamental Rights of the European Union (the 'Charter') and Art. 16(1) of the Treaty on the Functioning of the European Union (TFEU). The general position of the regulations on consent to data processing it that it should be given by a clear affirmative act establishing a freely given, specific, informed and unambiguous indication of the data subject's agreement to the processing of personal data relating to him or her (reg 32). Silence, inactivity and not unclicking pre-ticked boxes are specifically referred to as not constituting consent (reg 32).

The regulations also specifically provide for processing data in those scientific research contexts where the purpose of doing so is not fully identified. Reg. 33 stipulates that in this context 'data subjects should be allowed to give their consent to certain areas of scientific research when in keeping with recognized ethical standards for scientific research.' This would appear to be an endorsement of the idea that consent can have a degree of breadth in this context but equally a rejection of the idea that it can be totally general/open. Even then, this is subject to the proviso that data subjects should have the opportunity to tailor it to only certain areas of research or parts of research projects to the extent allowed by the intended purpose. The 'Pacman' tendencies shown by the research biobanking sector towards the use of human material and related data will have to be curbed in this light in the European context, a fact that has not been lost within related discourse, some of which was already pushing forward overly biobanking friendly interpretations of the regulations and angling for their reform even before they came into force (Hallinan & Friedewald, 2015). These curbs may end up supplementing and strengthening preexisting controls on secondary use of tissue and related data that exist within many member states. Such controls vary significantly across jurisdictions in terms of both their design and effects (see further Kaye et al., 2016). However, the following examples at least give a flavour of the overall landscape.

Norway has the helpful explicit starting point requirement that when previously collected material and data are subject to a different, wider or new use, that does not fall within the scope of the original consent, new voluntary, express and informed consent must be obtained (The Health Research Act). However, this does not apply where a Research Ethics Committee decides that the research in question is of significant interest to society and the data subject's welfare and integrity are ensured (section 35) and can also be disapplied by the Regional Committee for Medical and Health Research Ethics (REC) where it is not possible or very difficult to obtain new consent (section 15). Furthermore, uses of anonymized material do not require consent (section 20).

Finland allows the transfer of old clinical samples and research samples into a biobank following a special procedure notifying individuals and providing an opt-out to their inclusion (Biobank Act 2013, section 11). It allows non-consensual secondary use of materials taken from the dead subject to favourable ethical review and, in certain cases, permission from the National Supervisory Authority for Welfare and Health (section 20 of The Act on the Medical Use of Human Organs and Tissues).

The Netherlands operates a presumed consent approach to secondary use under its Civil Code where the resulting data is not capable of being 'traced back to the person from whom they originated' (Art. 7:467).

England and Wales permit secondary use of material for research and storage to that end (Human Tissue Act 2004, sections 1(1)(d)–(f)) to be carried out with appropriate consent. As seen in relation to removal, the concept of 'appropriate consent' is structured to allow ethically dubious scope for the use of proxy consent in relation to the dead and competent non-objecting children. It also provides for consent to the aforementioned activities to be deemed in relation to adults who lack the capacity to decide and have no valid advance decision to consent or refuse in force, the proviso being that the activity in question must be consistent with their interests and meet other requirements laid down in regulations (section 6(1) in conjunction with the Human Tissue Act 2004 (Persons who Lack Capacity to Consent and Transplants) Regulations 2006, particularly Reg. 3). Furthermore, the need for consent is dispensed with altogether for storage for the purpose of, and actual use of material for the purpose of, research in connection with disorders, or the functioning, of the human body where the material has come from the body of a living person provided it is carried out in circumstances where the person carrying it out is not in possession, and not likely to come into possession, of information from which the person from whose body the material has come can be identified and so long as it is part of research that is ethically approved in accordance with regulations made by the Secretary of State (Human Tissue Act 2004 sections 1[7]–1[9]).

As a whole the HTA was meant to embody an ethic of respect in the tissue research context but this goal was subverted (see further Figure 6.3).

The well documented context in which the reform of the law relating to human tissue in England and Wales was considered around the turn of the millennium was the public scandal that has erupted when it became common knowledge that practices of non-consensual taking, retention and use of body materials for research and certain other purposes that had been occurring for decades on a mass scale across England and other parts of the UK. In the run-up to passing a new Human Tissue Act in 2004 the Government committed itself to the principle of consent when dealing with human tissue both within its report *Human Bodies, Human Choices* (DoH, 2002) and in its *Proposals for new legislation on human organs and tissue* (DoH, 2003, 2). However, in the wake of medical lobbying this commitment was watered down. Echoing the NHANES III situation in the US, a less than fulsome approach to consent in relation to secondary use specifically was predicated partly on the expedience of saving money - with it being estimated at the time that the Act was passed that spending one minute to obtain consent for each of the relevant tissue samples obtained by the NHS each year would require an extra 1,339 full-time staff (Furness and Sullivan, 2004).

Figure 6.3 How vested interests nobbled the respect agenda during the creation of the HTA

In conclusion, whilst it may be true on a certain level that consent has been and remains contested in the tissue research context (Caulfield & Murdoch, 2017), it must be also understood that this has essentially been caused by an unwillingness to properly apply health law norms concerning it to this specific area. This, in turn, has been predicated not on any sound ethical principle but rather the expediency of wanting to facilitate greater and cheaper research use of tissue and with the effect that those used have less control over what happens to their material and related data.

6.6. Protecting communities and humans as a species

Earlier in this chapter, I argued that the traditional individualistic emphasis on liberty is too narrow to provide appropriate constraint in the human tissue research context. The reason for this is not only that it does not approach the interests of individuals in a rounded way but also because it fails to account for the relevance of communities and humans as a species. A community can be considered to have potential value in its own right as a kind of organic entity with a golden thread that extends from its past into its present and is ready to be woven into its future. It may incorporate both commonality and diversity across such factors as genetics, language, culture, ways of thinking about life, experience and so on. Meanwhile, a species is something further still – an element of the marvel of the diversity and magnificence of creation and something to be protected and valued in its own form.

The worth of both has increasingly been recognized within human rights frameworks. In the health context specifically, perhaps the leading role in this regard has been played by UNESCO through the auspices of which the international community has developed the Universal Declaration on the Human Genome and Human Rights (UDHGHR) and the Universal Declaration on Bioethics and Human Rights (UDBHR). The latter is a broad-based framework for control of scientific intervention as it affects human beings is particularly significant. Its preamble recognizes, amongst other things, the need for science and research to align with promoting 'the welfare of individuals, families, groups or communities, and humankind as a whole in the recognition of the dignity of the human person and universal respect for, and observance of, human rights and fundamental freedoms.' This is reflected in its Art. 2 aims including both the importance of human dignity generally and the specific need to protect current and future generations. To some extent, it is also evident in more specifically targeted provisions. For example, its provision for consent the Declaration stresses that in 'appropriate cases of research carried out on a group of persons or a community, additional agreement of the legal representatives of the group or community concerned may be sought' adding that this would be supplemental to rather than a substitute for individual informed consent (Art. 6(3)). There is also an emphasis on accounting for special vulnerability not just in individuals but also groups (Art. 8) and an emphasis on the need for life sciences to account for future generations (Art. 16).

The imperative of strong emphasis on protecting community and humans as a species has yet to be properly understood and embraced in the human tissue research context. Perhaps one of the most interesting and important case studies which highlights this is the research Arizona State University (ASU) conducted with the Native American Havasupai tribe. The Havasupai inhabit a vast reservation at the bottom of the Grand Canyon. In 1989, they sought help from John Martin, an anthropologist at ASU, to look into the causes of high incidence of Type 2 diabetes within their population. This was generating severe health issues for the tribe and was a leading cause of death amongst their adults.

Martin and two other university experts, nutritionist Professor Linda Vaughan and geneticist Professor Therese Ann Markow, worked up a proposal for a research project with Markow as lead. The upshot of this was that in 1990 the Havasupai Tribal Council were pitched and agreed to a project that would examine good nutrition and prevention in relation to Type 2 diabetes and use tests to determine tribal members susceptible to it and whether there was a genetic cause for that susceptibility. Participants for the first blood draw were verbally pitched the same story and due to limited capacities in English comprehension would doubtless have relied on that description of it rather than picked up the fact that the consent form that they were asked to sign referred to the project in incredibly broader terms as being a study of 'the causes of behavioural/medical disorders' (Hart & Sobraske, 2003, 50–52, 58).

The Hart Report – resulting from an independent investigation of the project commissioned by the ASU Board of Regents in 2003 after it became the subject of public scandal – found that Markow had specifically directed that the consent form should be so constructed, having already formed the intention to surreptitiously study the incidence of schizophrenia within the tribe (Hart & Sobraske, 2003, 58). The fact that she was wanted to study schizophrenia was a particular problem because the Havasupai considered it off limits because they took issue with the offensive underlying presumption that there was a high incidence of this condition amongst their populace and with the fact that this presumption may have originated with the depiction of one of their nineteenth century spiritual leaders as insane. She had been told of the sensitivities by Martin.

In a plosOrg blog, Lewis (2013) asserts that Markow never actually studied schizophrenia. She does not dispute the finding of the Hart Report that Markow secured funds from the National Association for Research on Schizophrenia and Depression before the project started and used some of that funding to pay for a first blood draw (Hart & Sobraske, 2003, 239), but questions whether it is relevant in the light of the fact that disease-specific associations sometimes fund basic research. However, this line of thinking is a red herring when one considers that Markow was found by the investigation to have, amongst other things, directed Kevin Zuerlein, the young psychiatrist she had appointed to conduct the blood draws, to secretly scan the medical files in the tribal clinic for records of psychiatric distress (Hart & Sobraske, 2003, 239); sought and gained approval from the ASU Institutional Review Board for research on it after the project had already started (Hart & Sobraske, 2003, 24); and described the project as being focused on diseases affected the Havasupai in general when interviewed about it (Hart & Sobraske, 2003, 132, 136–137, 152).

Ultimately, it is also clear that Markow led the project into further taboo areas as well. In 1993, she and Martin published a paper on inbreeding within the tribe (Markow & Martin, 2003). Then, when she moved to Arizona University, she took samples with her and collaborated with new colleagues in using them to study migration in relation to the tribe – a study which eventuated in a theory that the tribe had migrated from Siberia to North America (Karafet et al., 1997, 1999; Zegura et al., 2004) which was deeply disturbing and disruptive to its origin story and related spiritual and cultural beliefs and practices which centred partly on the notion that its people first manifested in the Grand Canyon. Markow also sent some of the samples to researchers from other universities leaving them free to explore their own further research avenues, despite the fact that the written consent form had stipulated that no information on the Havasupai would leave ASU. When later confronted about this, Markow used the fact that the samples were coded and individual donors could not be identified to disingenuously insist that no information had left ASU (Hart & Sobraske, 2003, 138). By the time the Hart Report was published, researchers from around the USA had already produced at least 23 scholarly articles on the topic, 17 of which were on the taboo subjects of schizophrenia, inbreeding and migration (Hart & Sobraske, 2003, 70–145).

Since the whole sorry saga was the subject of a legal settlement, no definitive conclusion may be reached on what, if any, infractions of the law occurred. However, notwithstanding what they signed, participants in this project did not know the purpose of what they were consenting to, and accordingly their consent would be classed as invalid (see discussion earlier in the chapter). Ethically speaking, the researchers not only breached the trust of those participants but the whole community and the Tribal Council. Not only was this breach self-evidently serious, but it was aggravated in intensity by the fact that it was grounded on the exploitation of the relationship that Martin had built up with the community after living in it for a year during the 1960s and subsequently writing extensively on its traditions and customs (Hart & Sobraske, 2003, 153–154).

Like other Native American Tribes, the Havasupai had learned by experience to by wary of the intentions of outsiders. It was because of the involvement of Martin that the Tribal Council developed a degree of confidence that they would be treated respectfully. This confidence in turn formed the basis of the Council's decision to allow the project to proceed and to encourage tribal members to participate, in spite of the fact that participation included blood draws which were in conflict with Havasupai beliefs about the need to die whole. As Kristof Van Assche, Sigrid Sterckx and Serge Gutwith point out in their thoughtful piece on the affair, only Christopher Armstrong, Markow's doctoral assistant, appeared to have moral qualms about how the samples and data were being used and whether this was authorized (2013, 59). On numerous occasions in 1996 and 1997, he stated to Markow that she was guilty of research misconduct, but when he notified ASU officials of his concerns, he was warned by ASU's lawyers that he was making serious and defamatory allegations (Hart & Sobraske, 2003, 28–29). No action was taken to scrutinize the weaknesses of her conduct, and the project as a whole until the tribe itself started to become aware of it 6 years later.

In addition to violating trust, the project illustrates how a range of other harms can be caused to a community through poorly conducted tissue research. Psychosocial harm can be caused to both individuals within a community and the community as a whole when research performed with it transgresses agreed limits and is specifically exacerbated when information is released that is stigmatizing or upsetting to its members. This can be worse where that community is based on a particular ethnic group, especially – as Van Assche et al. (2013, 67) note where the information relates to 'a psychiatric condition, or a socially unacceptable practice like inbreeding.' Because of the way it was conducted, the project with the Havasupai ended up being an exemplar – reinforcing 'the racial stereotype that Native Americans are unusually susceptible to certain types of disease' (Van Assche et al., 2013, 67). Such stereotypes are not only degrading and devaluing of a community but can also generate economic harm to it and its individual members by, for example, affecting their ability to obtain financial products and services such as loans and insurances. Citing the work of Tsosie (2007), they added that a community can also suffer cultural harm, which can eventually lead to its disruption (Van Assche et al., 2013, 67). Such harm was generated in the research project with the Havasupai because of the ways in which its focus of taboo issues disrupted the understandings and representations of the self within the community. Van Assche et al., argue that this was most extremely the case with the migration aspect of the project because it severely undermined the Havasupai's historical narrative and their sense of themselves and of their community (2013, 67).

6.7. Reciprocity and commodification

Towards the end of Chapter 1, I made the case that protecting humanity from instrumentalization or thingification is a necessary element of protecting its worth. In the context of exploring interpretations of Kant's work, I argued that this meant acting against exploitation in general not just its most extreme forms. I concluded that all forms of exploitation entail thingification and in this sense commodification because they promote the use value of subjects at the cost of their inherent value. However, I also noted that the term 'commodification' is also often used more selectively to describe thingification via money or money's worth being provided for the use of bodies and body parts at a level that goes beyond legitimate defrayment or reimbursement of reasonable actual expenses to constitute a potential inducement or, in some cases, even a source of profit. Efforts to constrain this in the research context have been relatively weak in two important respects.

Firstly, fixed sums of money and/or money's worth can be paid to participants which by their nature have the capacity to exceed expenses that have been or will be incurred. Notably, some empirical evidence suggests that even relatively small fixed sums can operate as an inducement – for example, $20 USD to participate in a genomics study in the US (McCarty et al., 2007). In the discourse of tissue research and to a lesser extent the discourse on transplantation, one finds emphasis on bio-rights, justice and reciprocity being used to justify such payments. It is, however, quite possible to construe these emphases not as providing a legitimate basis for commodification but simply as providing overlapping and mutually reinforcing bases for better protection of prospective subjects of tissue research and thus as capable of being useful tools for the protection of human worth. They are already often used to this end – for example, in the form of arguments that those who donate material ought to be informed of what is happening with the research that relates to that material and more specifically of any significant actual or potential health issue for them which is revealed by it.

Secondly, those receiving the material and/or further downstream in the process generally have considerable freedom to gain commercially from its use, in particular from:

(1) Discoveries about the tissues themselves or about the diseases and/or people that relate to them;
(2) Products and therapies built off the back of these discoveries; *and*
(3) Products built partly out of the materials themselves.

Even health-related global human rights provisions which are designed to prevent commodification in general do little to prevent the above forms of gain. The CHRB, for example, opposes financial gain being made from the body and its parts but is silent about the broader commodification implicit in the first and second and specifically treats the third as an exception to its scope by stressing that it does not apply to issues relating to the patentability and patenting of biotechnological inventions (Explanatory Report, para. 134) or prohibit the sale of a medical device incorporating human tissue which has been subjected to a manufacturing process as long as the tissue is not sold as such (Explanatory Report, para. 131). A logical and necessary corollary of adhering to the full spirit of opposing commodification would be to prevent all three by at the very least compelling engagement in tissue research to be on a not-for-profit basis.

6.8. Conclusions

Human tissue research has been extensively hyped and allowed to occur in ways and under conditions that are not productive. From a respect for worth based perspective, both it and its exploitation should only occur on a not for profit ethos and serve appropriate ends, including a holism in the medical context. Furthermore, in so far as it is needed that need must cease to be treated as a justification for the conditions of its use to deviate from full protection of the interests of 'donors' and their wider communities. Amongst other things, this will necessitate consent in this field being handled in a way that aligns with how it is normatively treated in the medical context more generally.

Whilst some commentators regard obtaining specific and informed consent as problematic in the human tissue research context, the problem can actually be categorized as being simply that some want it to be practiced free of some of the constraints that flow from the need to obtain it. This is evident, for example, when one looks at the four problems that van der Graaf and van Delden cite with respect to obtaining such consent in the research biobanking context specifically (2011, 15):

- *Firstly*, they express concern about requiring specific consent in the context of an activity that entails storage for as yet unspecified research purposes when from a respect for worth point of view it is clear that if specific consent is needed to protect interests then it should be required.
- *Secondly*, they express concerns about needing consent in the context of so-called leftovers when actually this amounts to nothing more than a minor cost and convenience issue.
- *Thirdly*, they express clearly misplaced concerns about whether recontacting participants to gain further consent contributes to respect for their autonomy.
- *Finally*, they express concerns that allowing withdrawal of consent can diminish the quality of research when clearly providing for it is central to being accountable to the autonomy of participants.

Some authors have tried a different tack of arguing that there is not the same ethical imperative for specific and informed consent in the human tissue research context as there is elsewhere. For example, in relation to the processing of data specifically, Hallinan and Friedewald argue that this is so because, unlike contexts like social networking and e-commerce, the relationship between those whose data is being processed and the data controller is not adversarial (Hallinan & Friedewald, 2015, 20). Indeed, they go as far as supporting Laurie's argument that there is a confluence of interests between the parties (Laurie, 2002, 165–167) and stress that the data processing is not aimed at any individual or done with intent to have any effect on the individual (Hallinan & Friedewald, 2015, 20). However, the first of these arguments may be criticized as naïve given that the values and interests that each party comes to the relationship with are likely to be different from the outset and further diverge over time and given that those carrying out the research are invariably either using it and/or allowing others to use it for commercial purposes. Furthermore, both the first and the second arguments ignore the fact that what matters above what one is aiming or intending in doing research is the actual harms and risks that one is imposing by doing it.

The law's emphasis on specific and informed consent can thus be conclude to be vital. However, protecting 'donors' and their wider communities in this context should also be seen as necessitating the law also being developed in new directions. One of the reforms that has been mooted is the development of a genomic tort to enable people to control the flow of genetic information about themselves, protecting the empowerment aspect of their dignity in the process. Laurie (2002) and Brownsword (2003) have argued that a legal right to such control could emerge as part of a broad concept of privacy or from a proprietary interest that individuals might have in their genetic information. However, as van Assche et al., have pointed out, 'accepting proprietary rights in relation to genetic information is likely to entail recognition of proprietary rights to the very tissue or samples that hold genetic information' (2013, 83).

This is an outcome that I earlier described as fuelling commodificatory thought and thus ultimately hard to reconcile with a goal of strengthening a culture of protection of dignity or worth. Accordingly, I would agree with van Assche et al., that the privacy angle may be more promising (2013, 83). Either outcome could, they suggest, cover a wide range of problems, including those suffered by the Havasupai tribe (2013, 83). However, this perhaps ignores the fact that it would not cover rights of control over use of tissue, which as I have illustrated in this chapter, and is also specifically evident in the Havasupai case can present concerns in its own right that are not entirely reducible to issues of the control over information stemming from its use.

Furthermore, as also emphasized in this chapter, the use of human tissue and biobanking more generally gives rise not just to considerations of control flowing from concern to protect autonomy or the autonomy aspect of dignity or worth but also issues of protection – and hence constraint. Most especially this is so with classes lacking the ability to make the relevant decisions, but it also so more generally with respect, for example, to uses involving detriment to the source of the material or commodification. In this regard, it may be preferable to go down another route of directly relying on dignity to protect interests in this context. Many jurisdictions already have one or more routes through which they can do this. As we have seen, for example, many jurisdictions protect dignity specifically within their domestic frameworks. and some incorporate one or more dignity protecting global or regional instruments into their domestic law.

An alternative strategy that could bring a more systematic and proactive approach to reform would be for it to be addressed at the global level. Chen and Pang have advocated this as a strategy to help improve governance in lower- and middle-income countries in particular and to foster proper regulation of cross border sharing of benefits and data (2015, 113). They suggest that a global governance framework ought to include the following six elements:

(1) Respecting participants and donors of biological samples and protecting their privacy and confidentiality;
(2) Informing participants and donors of potential risks through initial consultations;
(3) Sharing samples, data and benefits in a fair, transparent and equitable manner;
(4) Ensuring quality and interoperability of samples and their associated data;

(5) Improving public awareness, trust and participation in biobanks; and
(6) Defining the role of the private sector in the use of knowledge derived from biobank operations.

These elements would undoubtedly be useful, but I would argue that further work on global governance in this area ought to be driven from a more explicitly human rights based framework and that it should thus derive from further work by organizations like the Council of Europe and UNESCO.

References

Allen, C., et al. (2013) Data sharing, biobanks and informed consent: a research paradox. *McGill JL & Health.* 7, 85.

Andrews, L. & Nelkin, D. (2001) *Body Bazaar.* New York, NY: Crown Publishers.

Árnason, G. (2007) *Icelandic Biobank: A Report for GenBenefit.* Available at www.uclan.ac.uk/genbenefit. Accessed 1.3.2019.

Barilan, M. (2011) The biomedical uses of the body: Lessons from the history of human rights and dignity. In Lenk, C., Hoppe, N., Beier, K. & Wiesemann, C. (eds.) *Human Tissue Research: A European Perspective on the Ethical and Legal Challenges.* Oxford: Oxford University Press, pp. 3–14.

BCC. (2011) *Global Market for Biobanking to Surpass $183 Billion in 2015.* Available at www.bccresearch.com.

Beresford, E. (1968) *The Wombles.* London: Ernest Benn Ltd.

Bister, M. (2011) Discovering informed consent: A case study of the practices of informed consent to tissue donation in Austria. In Lenk, C., Hoppe, N., Beier, K. & Wiesemann, C. (eds.) *Human Tissue Research: A European Perspective on the Ethical and Legal Challenges.* Oxford: Oxford University Press, pp. 169–177.

Brownsword, R. (2003) An interest in human dignity as the basis for genomic torts. *Washburn Law Journal.* 42(3), 413.

Budmir, D., et al. (2011) Ethical aspects of human biobanks: A systematic review. *Croatian Medical Journal.* 52(3), 262–279.

Caulfield, T. & Murdoch, B. (2017) Genes, cells and biobanks: Yes there's still a consent problem. *PLoS Biology.* July, 15(7), e2002654.

Chalmers, D., et al. (2016) Has the biobank bubble burst? Withstanding the challenges for sustainable biobanking in the digital era. *BMC Medical Ethics.* 17, 39.

Chen, H. & Pan, T. (2015) A call for global governance of biobanks. *Bull World Health Organ.* 1 February, 93(2), 113–117.

Department of Health. (2002) *Human Bodies: Human Choices.* London: DoH.

Department of Health. (2003) *Proposals for New Legislation on Human Organs and Tissue.* London: DoH.

De Souza, Y. & Greenspan, J. (2013) Biobanking past, present and future: Responsibilities and benefits. *AIDS.* 27(3), 303–312.

De Vries, R., et al. (2016) Understanding the public's reservations about broad consent and study-by-study consent for donations to a biobank: Results of a national survey. *PloS ONE.* 11, 7 e0159113. Available at doi: 10.1371/journal.pone.0159113.

Eriksson, G. & Helgesson, L. (2005) The right to withdraw consent to research on biobank samples. *Medicine Health Care and Philosophy.* 8(3), 315–321.

Garrison, N., et al. (2015) A systematic literature review of individuals' perspectives on broad consent and data sharing in the United States. *Genetics in Medicine.* 19 November, 18(7), 663–671. Available at doi: 10.1038/gim.2015.138.

Garwood-Gowers, A. (2005) The proper limits for medical intervention that harms the therapeutic interests of interests of incompetents. In Garwood-Gowers, A., Wheat, K. & Tingle, J. (eds.) *Contemporary Issues in Healthcare Law and Ethics.* Amsterdam: Elsevier, pp. 191–212.

Garwood-Gowers, A. (2011) Respect as a precondition for use of human tissue for research purposes. In Lenk, C., Hoppe, N., Beier, K. & Wiesemann, C. (eds.) *Human Tissue Research: A European Perspective on the Ethical and Legal Challenges.* Oxford: Oxford University Press, pp. 25–34.

Garwood-Gowers, A. (2013) Time to address the problem of post-mortem procurement of organs for transplantation occurring without proper pre-mortem consent. *European Journal of Health Law.* 20, 383–408.

Garwood-Gowers, A. & Olsena, S. (2017) Informed consent. In den Exter, A. (ed.) *European Health Law.* Leiden: Brill, pp. 245–272.

Hallinan, D. & Friedewald, M. (2015) Open consent, biobanking and data protection law: Can open consent be 'informed' under the forthcoming data protection regulation? *Life Sciences, Society and Policy.* 24 January, 11(1), 1.

Hansson, M. (2005) Building on relationships of trust in biobank research. *Journal of Medical Ethics.* 31, 415–418.

Hart, S. & Sobraske, K. (2003) *Investigative Report Concerning the Medical Genetics Project at Havasupai*, 23 December. Available at http://cnhp.montana.edu/conference/HartReport.pdf.

Henderson, G., et al. (2013) Characterizing biobank organizations in the US: Results from a national survey. *Genome Medicine.* 25 January, 5(1), 3. Available at doi: 10.1186/gm407.

Hoeyer, K. (2005) The role of ethics in commercial genetic research: Notes on the notion of commodification. *Medical Anthropology.* 24, 45–70.

Hoeyer, K. (2008) The ethics of research biobanking: A critical review of the literature. *Biotechnology and Genetic Engineering Reviews.* 25(1), 429–452.

Karafet, T., et al. (1997) Y chromosome markers and trans-Bering strait dispersals. *American Journal of Physical Anthropology.* 102(3), 301.

Karafet, T., et al. (1999) Ancestral Asian source(s) of new world y-chromosome founder haplotypes. *American Journal of Human Genetics.* 64(3), 817.

Kaye, J., et al. (2016) Consent for biobanking: The legal frameworks of countries in the BioSHaRE-EU Project. *Biopreservation and Biobanking.* June, 14(3). Available at https://doi.org/10.1089/bio.2015.0123.

Kirchhoffer, D. & Dierickz, K. (2012) Human dignity and consent in research biobanking. *The South African Journal of Bioethics and Law.* 5(2), 74–77.

Lassek, A. (1958) *Human Dissection: Its Drama and Struggle.* Springfield, IL: Charles C. Thomas.

Laurie, G. (2002) *Genetic Privacy: A Challenge to Medico-Legal Norms.* Cambridge: Cambridge University Press.

Lenk, C., et al. (2011) Preface. In Lenk, C., Hoppe, N., Beier, K. & Wiesemann, C. (eds.) *Human Tissue Research: A European Perspective on the Ethical and Legal Challenges.* Oxford: Oxford University Press, pp. IX-XXII.

Lewis, R. (2013) *Is the Havasupai Indian Case a Fairy Tale?* Available at https://blogs.plos.org/dnascience/2013/08/15/is-the-havasupai-indian-case-a-fairy-tale/.

Lipworth, W., et al. (2011) Tissue donation to biobanks: A review of sociological studies. *Sociology of Health and Illness*. 33(5), 792–811.

Markow, T. & Martin, J. (1993) Inbreeding and developmental stability in a small human population. *Annals of Human Biology*. 20(4), 389.

Matthews, P. (1995) The man of property. *Medical Law Review*. 3, 251.

McCarty, C. (2007) Informed consent and subject motivation to participate in a large, population-based genomics study: The Marshfield Clinic Personalized Medicine Research Project. *Community Genetics*. 10(1), 2–9.

Meir, K., et al. (2014) Biobank networking for dissemination of data and resources: An overview. *Journal of Biorepository Science for Applied Medicine*. 2, 29–42.

Murphy, J., et al. (2009) Public perspectives on informed consent for biobanking. *American Journal of Public Health*. December, 99(12), 2128–2134. Available at doi: 10.2105/AJPH.2008.157099.

NIH. (2014) *Genomic Data Sharing Policy, 79 Fed.* Reg. 51345–54. 28 August.

The Nuffield Council on Bioethics. (1995) *Human Tissue: Ethical and Legal Issues*. Available at http://nuffieldbioethics.org/wp-content/uploads/2014/07/Human-tissue.pdf.

Ogbogu, U., et al. (2014) Newspaper coverage of biobanks. *Peer Journal*. 31 July, 2, e500. Available at doi: 10.7717/peerj.500.

Simon, C., et al. (2011) Active choice but not too active: Public perspectives on biobank consent models. *Genetics in Medicine*. 1 September, 13(9), 821–831. Available at doi: 10.1097/GIM.0b013e31821d2f88.

Tsosie, R. (2007) Cultural challenges to biotechnology: Native American Genetic resources and the concept of cultural harm. *Journal of Law, Medicine and Ethics*. 35(3), 396.

Van Assche, K., et al. (2013) Protecting dignitary interests of biobank research participants: Lessons from Havasupai Tribe v Arizona Board of Regents. *LIT*. 5(1), 55–85.

Van der Graaf, R. & van Delden, J. (2011) Exploring an alternative to consent in biobank research. In Lenk, C., Hoppe, N., Beier, K. & Wiesemann, C. (eds.) *Human Tissue Research: A European Perspective on the Ethical and Legal Challenges*. Oxford: Oxford University Press, pp. 15–24.

Vaught, J., et al. (2011) Biobankonomics: Developing a sustainable business model approach for the formation of a human tissue biobank. *Journal of the National Cancer Institute*. 1 January, 42, 24–31.

Waldby, C. & Mitchell, R. (2006) *Tissue Economies: Blood Organs and Cell Lines in Late Capitalism*. London: Duke University Press.

Weir, R. & Olick, R. (2004) *The Stored Tissue Issue: Biomedical Research, Ethics and Law in the Era of Genomic Medicine*. Oxford: Oxford University Press.

Wendler, D. (2006) One-time general consent for research in biological samples. *British Medical Journal*. 332, 544–547.

Zegura, S., et al. (2004) High-resolution SNPs and microsatellite haplotypes point to a single, recent entry of Native American y chromosomes into the Americas. *Molecular Biology and Evolution*. 21(1), 164.

Conclusion

There is a clear need for a theoretical lens when addressing questions of medical ethics generally and medical use specifically

The idea that one can understand and address specific ethical questions in medicine without a theoretical lens for doing so is popular in medicine, often taking the form of support for principlism. The so-called Four Principles Approach developed by Beauchamp and Childress (2013) is the most well-known and used form of principlism in bioethics (Lawrence, 2007, 34). It emphasizes autonomy, justice, beneficence and non-maleficence. Brannmark (2017) has recently developed a five principles approach which emphasizes autonomy, dignity, integrity, privacy and protection in the context of vulnerability.

The problem that strikes at the heart of all forms of principlism is that they are anti-foundational (Meilaender, 1995). In particular, they lack a basis upon which to resolve conflicts between the principles that they emphasize. These conflicts are well known to be commonplace in medical practice and some of them have been highlighted in the course of this book, including between autonomy on the one hand and beneficence, non-maleficence, dignity and protection in the context of vulnerability on the other. Part of the proper role of medical ethics should be to identify and use a suitable theoretical lens to understand and address these conflicts.

Brannmark (2017) rejects such an exercise based on his concern that it would damage pluralism to develop a commitment to one or more absolutes. However, to endorse the relativistic notion that there are no absolutes is itself a form of absolutism. As Cooper has stressed, a view or theory must not simply be something that comes out of the mouth of some theorist but also liveable (2002, 13). As with determinism, discussed in Chapter 1, relativism is not liveable because to hold to it is self-contradictory. It is also pernicious. Embracing it does not actually eliminate the need to form judgments on ethical issues. Rather, it simply leaves those judgments to be made on the frontline where they will tend to be made through the lens of whatever mixture of unconsciously as well as consciously held values and beliefs those making them have. In this sense, taking a relativistic approach to medical ethics, based on principlism or otherwise, would lead to a weakening of control over medical power in the real world and leave prospective subjects of medical use more vulnerable to exploitation.

Pluralism would be much better protected by embracing the inevitable need for a theoretical lens and agitating for one that is influenced by the concerns of pluralism. In terms of substance, the respect for worth based approach taken in this book fits this bill of being an approach that could be put forward. It is highly protective of the principle of leaving individuals free to live their own vision of the good life, notwithstanding that it departs from libertarianism in some ways. In terms of process, it is about considering and selecting between possible approaches through an open and ideally at least somewhat democratic process. Many forms of principlism do not even have a good process for selecting principles. Brannmark (2017) selected what he considers to be the central relevant principles arising within general and health specific human rights instruments. This is not a democratic process of selection per se but at least has some democratic legitimacy since these are instruments which are both relevant and designed largely around the objective of constraining excess. By contrast, Beauchamp and Childress simply derived their principles from what they observed was already being used in medical decision-making practice. In other words, the process simply held up a mirror to reflect back the status quo. Furthermore, the status quo entails built in support for vast discretionary power, since there are many issues which can be addressed in more than one way using the principles. Such discretionary power would be particularly problematic in the context of medical use, since this is an area where those working on the frontline have in certain respects a vested interest to apply discretion in ways that bring about use, rather than embody a neutral and well considered reflection on the prospective subjects interests.

The preference for respect as the preferred theoretical lens was entirely justified

An inclusive respect for worth based approach to viewing questions of ethics generally and medical use of the human being specifically, follows on logically from understanding the nature and purpose of reality in a holistic fashion. Whilst personhood and dignity have been used extensively in Western discourse as proxies for worth, both are problematic. Both can and have been deployed in support not just of a notion of worth that does not extend fully or even at all beyond human species but also even selective as between which members of the species to protect. Amongst other things, this has meant that they do not provide an adequate ideological bulwark against the kinds of both ideologically and opportunistically grounded exploitation of vulnerability that frequently happens in the context of medical use as I exposed in detail in Part II of this book.

Towards the end of Chapter 1, I laid out what I consider to be the preconditions for medical use that flow from respect for human worth. I made the case that it should be productive in terms of the ends it pursues and compatible with four norms of:

(1) Protecting the equal value of all members of the species;
(2) Protecting individuals, groups and the species from instrumentalization;

(3) Protecting against fundamental interferences with human nature and form; *and*
(4) Providing appropriate protection of other relevant values, such as autonomy and beneficence.

There are a number of reasons to support this schema:

- *Firstly*, even if the requirement for productivity is not self-evident, it is pretty obvious and uncontentious. Only the most ardent of libertarians would think it acceptable to proceed with medical use that was destructive, counterproductive or even merely unproductive.
- *Secondly*, the four norms broadly ally with what most dignitarians who support the notion of inherent worth would come up with if they were developing a schema. The very notion of dignity as inherent worth implies a norm of protecting all members of the species as of equal value and also a norm of not permitting the mere means to an end or instrumental use of the human being. Furthermore, most dignitarians also treat it as placing limits on what can be done with human nature and form. For example, dignitarian opposition to practices like cloning and other forms of genetic modification is commonplace and is reflected in a number of health related human rights instruments as seen in Chapter 3. Finally, the fourth norm is simply a recognition that there appear to be no other absolutes implied by dignity or inherent worth but that there are, nonetheless, a number of other principles that can be relevant to it.
- *Thirdly*, the norms, and possibly the productivity requirement as well, gel nicely with a human rights law based approach to constraint of medical use. As was illustrated in Chapter 3, most general and health specific human rights instruments refer to dignity in such a way as to support the protection of human worth and tend at least to some degree to organize their schema for rights and protections accordingly. Most instruments have rights and protections that protect equality (albeit ones that typically do not extend fully to the not yet born and the dead) and rights and protections that ensure the body related interests of human beings can only be circumscribed by reference to the conflicting rights of others, not their mere (claimed) needs. Many of those that are focused on the health field also show at least some inclination to protect against fundamental interferences with human form – such as via earlier referred to tendencies to constrain practices like cloning and genetic engineering.
- *Fourthly*, the schema as a whole seems to be 'road worthy' when applied to the three case studies in Part II of the book. It certainly helped to frame and provide a language for the discussion of the extensive problems that arise in medical use related discourse, practice, governance and law.

The principal advantage of the schema is that it can be used not just as a means of addressing the ethics of specific forms of actual and proposed medical use in a tangible way but could also in the future be used as a template for reform of human rights instruments, particularly or at least in the health field. Its specificity stands as a rebuke to those who consider dignity – and perhaps by implication worth – to be empty of meaningful content and as a reasonable response to those who have legitimate concerns that it can be misused to all manner of mischievous liberty hating ends. We must have something particularly in an era where the threat of misuse of dignity in the biosciences is now starting to be exceeded by the threat of the misuse of freedom.

The benefits of a case study approach and the themes emerging from it

The case study approach adopted in this book has first and foremost exposed serious problems in the three areas studied. Warfare related medical use has cultivated and been cultivated by an emphasis, on authoritarianism, lack of openness, racist and xenophobic nationalism and an ideological and opportunistic willingness to exploit inherent and socially constructed vulnerabilities. Solid organ transplantation has cultivated and been cultivated by a counterproductively mechanistic and industrialized approach to medicine, a concomitant lack of emphasis on the fundamental issues raised by it and the value of alternatives to it. Even many who support reliance on it would accept Fox and Swazey's criticism that overall it has been pursued too ardently and hubristically (2017, xviii).

Finally, human tissue research has cultivated and been cultivated by the idea that it will bring about a golden revolution in medicine and beyond. This has contributed to a soporific acceptance of the ends pursued by it and added to the willingness to pursue these ends via means inconsistent with human interests.

Perhaps the other major benefit of taking a case study approach has been the way it has exposed issues which cut across areas and thereby revealed much about the problematic character of medical use as a whole. This is deeply significant. To date, context, class and theme specific ways of addressing medical use have formed the virtual entirety of literature surrounding it. This has enabled arguments for diluting protection made with respect to one context, class or theme to be presented as if they are particular and exceptional to it when, in fact, they tend to be more or less mirrored elsewhere. One example of this is the way that the utilitarian notion that the end can justify the means appears in all three areas covered by the case studies. Like a chameleon, it simply adapts its presentation to the particular context in which it is appearing. In warfare related medical use, it is typically bolstered by reference to the importance of the nation state. In relation to solid organ transplantation, it is bolstered by the (ultimately false) notion that there is a shortage of suitable materials for use which can only be overcome by extraordinary procurement measures, including measures that have profoundly adverse ethical implications, such as adopting questionable approaches to death and less than fully protective systems for so-called post-mortem procurement of solid organs. And in relation to human tissue research, it is bolstered by a hyped notion of the potential damage that would be caused by applying protective norms fully, such as in the context of thinking about consent. Another example is the way that libertarianism has been used as a lever to justify practices that would be classified as abusive from a respect for worth perspective, such as voluntary sale of body materials; broad and even one time general consent in the context of human tissue research, transplantation following voluntary euthanasia *and* transplant related removal of essential to life organs from people who have given advance consent to be pronounced dead at a point where they ought to be classified as alive (such as under the Pittsburgh Protocol).

One can also see by looking across contexts, that the class reach of interests are repeatedly being understood in an overly narrow way, such as to enable unjustifiable use. Classes who are already naturally vulnerable to misuse are rendered even more so by the application of personism. This has happened on a broad basis in the context of warfare related medical use and also on a narrower but still pernicious basis in the context of human tissue research and transplantation in relation to the dead. Another example is how the role of consent in serving human worth has been underplayed, such as via support for the notion of irrevocable consent in human tissue research and presumed consent in both human tissue research and transplantation.

Finally, problems with productivity can also be seen across the different case study areas of use. Human tissue research is much more mixed in value than is typically presented, and some forms of it are clearly counterproductive, transplantation is of doubtful value in general, and warfare related use has been shown to be in large part destructive. The counterproductively mechanistic orientation and industrialized character of establishment medicine is also reflected in all of the case study areas.

The combined impact of misuse across the three case study areas is such as to provide a clear sense of just how very problematic medical use is as a phenomenon and how drastically its practice will need to be reined in. At the moment, there is simply far too much credence given to the propagated value of many forms of medical use. Worse still, there is far too much willingness to let them be practiced at the cost of proper protection of those used. This will only change where we not only avoid liberty levered exploitation but also recognize that '(n) o pragmatic, utilitarian or otherwise consequential argument can override basic ethical principles' (Döring, 2014, 142).

References

Beauchamp, T. & Childress, J. (2013) *Principles of Biomedical Ethics* (7th Edition). Oxford: Oxford University Press.

Brannmark, J. (2017) Respect for persons in bioethics: Towards a human rights based account. *Human Rights Review*. 18, 171–187.

Cooper, D. (2002) *The Measure of Things: Humanism, Humility and Mystery*. Oxford: Clarendon Press.

Döring, O. (2014) Bioethics and exceptionalism: A German example of learning from medical atrocities. In Nie, J-B., et al. (eds.) *Japan's Wartime Medical Atrocities.* Abingdon: Routledge, pp. 141–154.

Fox, R. & Swazey, J. (2017) *Spare Parts: Organ Replacement in American Society.* Abingdon: Routledge.

Lawrence, D. (2007) The four principles of biomedical ethics: A Foundation for current bioethical debate. *Journal of Chiropractic Humanities.* 14, 34–40.

Meilaender, G. (1995) *Body, Soul and Bioethics.* Notre Dame: University of Notre Dame.

Index

Note: Page numbers in *italic* indicate a figure on the corresponding page